Soybean - Biomass, Yield and Productivity

Edited by Minobu Kasai

Published in London, United Kingdom

IntechOpen

Supporting open minds since 2005

Soybean - Biomass, Yield and Productivity
http://dx.doi.org/10.5772/intechopen.76784
Edited by Minobu Kasai

Contributors
Milton Edgar Pereira-Flores, Flavio Justino, Obed Mwenye, Rouxlene Van Der Marwe, Leon Van Rensburg, Angeline Van Biljon, Zhenhao Xi, Ling Zhao, Yahua Chen, Qiuyu Tang, Lanpeng Li, Chong Peng, Huizhi Gao, Qian Li, Kiyoshi Nagasuga, Heike Bücking, Arjun Kafle, Kevin Garcia, Vincent Peta, Jaya Yakha, Alex Soupir, Fabiola Araceli Guzmán-Ortiz, Rosalva Mora-Escobedo, Ciro Baruchs Muñoz Llandes, María Del Carmen Robles-Ramírez, Alma Delia Román-Gutiérrez, Javier Castro-Rosas, Thandiwe Nleya, Matthew Schutte, Kirstin Wurms, Annette Ah Chee, Paul Sutherland, Lorena Deleanu, Constantin Georgescu, George Catalin Cristea

First published in London, United Kingdom, 2019 by IntechOpen
IntechOpen is the global imprint of INTECHOPEN LIMITED, registered in England and Wales, registration number: 11086078, The Shard, 25th floor, 32 London Bridge Street
London, SE19SG - United Kingdom
Printed in Croatia

British Library Cataloguing-in-Publication Data
A catalogue record for this book is available from the British Library

Additional hard copies can be obtained from orders@intechopen.com

Soybean - Biomass, Yield and Productivity
Edited by Minobu Kasai
p. cm.
Print ISBN 978-1-78985-373-5
Online ISBN 978-1-78985-374-2

For EU product safety concerns:
IN TECH d.o.o., Prolaz Marije Krucifikse Kozulić 3, 51000 Rijeka, Croatia, info@intechopen.com or visit our website at intechopen.com.

Meet the editor

Minobu Kasai is a professor at Hirosaki University in Japan. He received his PhD degree from the University of Tokyo. His main research and papers are concerned with environmental physiology of higher plants, including soybean, and are mainly summarized in a chapter in A Comprehensive Survey of International Soybean Research (2013) and a mini-review in Trends in Photochemistry and Photobiology (2014).

Contents

Preface

Plants are important for a permanent ecosystem, because in the ecological pyramid plants support all the other living organisms at the base. Very important organization is thought to be the integral process of resource, transport, partitioning, metabolism, and production, which involves yield, biomass, and productivity in plants. Accordingly, it is important to obtain more information about the knowledge concerning yield, biomass, and productivity in plants.

Soybean is one of the main crops largely contributing to our life, which is thought to be connected to our ecosystem through the above-mentioned integral process. This book focuses on the soybean, and reviews and research concerning the yield, biomass, and productivity of soybean are presented herein. This text updates the book published in 2017.

Although there are many difficulties, the main aim of this book is to present a basis for the above-mentioned integral processes of resource, transport, partitioning, metabolism, and production, which involves yield, biomass, and productivity in plants (soybean), and to understand what supports this basis and the integral process.

It is hoped that this and the preceding book will be essential reads.

Minobu Kasai, PhD
Professor,
Department of Biology,
Faculty of Agriculture and Life Science,
Hirosaki University, Japan

Chapter 1

Fungicidal Activity of Soybean Oil against Powdery Mildew on Wheat

Kirstin V. Wurms, Annette Ah Chee and Paul Sutherland

Abstract

Products derived from soybean crops are not only global food staples, but are also used in pharmaceuticals, industry, and agriculture. Soybean oil (SBO) and other oils are often used as adjuvants in agricultural sprays to facilitate spread of the active ingredient (a.i.) across the plant surface. This chapter describes original research in which a natural fungicide (biofungicide) was formulated using SBO as the a.i. Antimicrobial activity of SBO against a powdery mildew (PM) pathogen, *Blumeria graminis* f. sp. *tritici*, was measured, as well as effects on plant health and yield of wheat plants. Results were compared with a conventional fungicide and another lipid biofungicide. The mode of action was investigated using scanning electron microscopy. Results showed that SBO provided PM control equal to the conventional fungicide when plants were adequately spaced and caused collapse of fungal structures and extrusion of cell contents. Commercialisation potential of SBO biofungicide is discussed.

Keywords: biofungicide, horticultural oils, powdery mildew, scanning electron microscopy, soybean oil, wheat

1. Introduction

Soybean (*Glycine max*), a species of legume native to East Asia, is economically the most important bean in the world, providing vegetable protein for millions of people and animals and ingredients for hundreds of chemical products including pharmaceuticals, cosmetics and biofuels [1]. Soybeans have a myriad of health benefits for humans including their ability to stimulate metabolism, promote heart health and osteotropic activity, protect against cancer, prevent birth defects, aid digestion, increase circulation and decrease the risk of diabetes [2], but in this chapter we focus on the use of soybean oil (SBO) in agriculture to improve plant health.

Soybean oil is a vegetable oil that is solvent-extracted from pressed seeds of soybean, followed by refinement, blending and optional hydrogenation, and is one of the world's most widely consumed edible plant oils [3, 4]. While plant-derived oils, such as SBO, are predominantly used in agriculture as an adjuvant to aid the spread of pesticides over plant surfaces and also to help the pesticide to stick to the plant surface [5, 6], SBO is also directly antimicrobial against a range of powdery mildew fungi [7–10], *Botrytis cinerea* [11] and bacteria such as *Staphylococcus aureus* and *Escherichia coli* [12]. In addition, SBO has insecticidal activity against mites [13], whitefly and aphids [14, 15] and insects associated with stored grain products [16] . However, there are few commercial products in horticulture that use plant oils as the active ingredient (the most notable exception being neem oil), with most spray oils comprising mineral oils that are refined from petroleum. Commercial development of SBO as a pesticide offers the advantages of reduced reliance on petroleum products, and the use of an edible oil is considered to be less toxic to human health and the environment. However, fats and oils are often associated with chlorosis and necrosis of plant tissue [13, 17–19], and other problems include inconsistent activity, handling and application difficulties, spoilage and development of unpleasant odours, and these issues need to be considered when developing a SBO fungicide.

Powdery mildew (PM) disease is characterised by fluffy white lesions on the surfaces of aerial plant tissues. It is caused by pathogens from the Erysiphales order and is responsible for significant yield losses globally in crops such as cucurbits, apples, roses, tomatoes, grapes and various cereals such as wheat and barley [20, 21]. PM is one of the most economically damaging plant diseases around the world. For instance, losses to barley PM in the State of Western Australia have been estimated at $30 million (AUD) annually [22]; losses account for approximately 15% of total crop revenue for American North Western hop growers, which equated to over $30 million USD in the year 2000 [23]; and the introduction of PM-resistant grape varieties into the State of California alone has been estimated to yield $48 million (USD) in annual cost savings [24]. There are serious limitations with existing PM control methods, such as pathogen resistance to demethylation inhibitor fungicides [25, 26]. Moreover, synthetic pesticide use has been clearly linked to human health concerns such as increased incidence of respiratory disease and cancer [27]. There are also limitations on the use of sulphur and copper-based fungicides, considered to be more natural fungicides, in organic systems because sulphur can act as a nose and eye irritant [28] and because heavy metals like copper accumulate in soils with intensive copper fungicide use over time, resulting in phytotoxicity [29]. These issues are driving the development of biofungicides (fungicides comprising biological control agents and/or natural products) that are suitable for both organic and conventional growers, considered safer in terms of human health and which provide an environmentally benign option for durable disease control. Furthermore, PM strains mutate and develop resistance rapidly to synthetic pesticides, but there are few documented instances of resistance development to oils [30, 31]. Biofungicides can be used as standalone products, or in integrated disease control programmes that combine treatments with multiple modes of action, to reduce the application number of traditional synthetic pesticides and to delay the onset of resistance.

The principal aim of this study was to investigate the potential of SBO as a biofungicide to control powdery mildew. To achieve this aim, SBO effects on disease control efficacy and plant health and yield were compared under regulated conditions found in a controlled environment room and a glasshouse versus the more variable conditions in a field situation. SBO performance was also compared versus conventional fungicides and another lipid biofungicide-emulsified anhydrous milk fat (AMF) from cows' milk, since SBO and AMF were the two top candidates from a preliminary study investigating lipid biofungicide action against PM [32]. Product mode of action (MoA) was also investigated using scanning electron microscopy (SEM), because knowledge of the MoA permits a product to be used more effectively in relation to timing and mode of application, and helps to manage the risk of target organisms developing resistance to the control product [33, 34]. MoA is also often necessary for product registration. Wheat (*Triticum aestivum*) was chosen as the ideal crop for this study because it is a global food staple for which PM is a common disease problem [35, 36] and because wheat plants can be easily and quickly grown. Given that the leaf surface of wheat is non-hairy and robust, it is also more likely to produce clear images in SEM following sample preparation by cold stage freezing and sputter coating with gold.

Findings obtained from the data are discussed with respect to the commercialisation potential of SBO biofungicide.

2. Controlled environment (CE) room and glasshouse trials

2.1 CE trial methods

PM-susceptible 'Endeavour' wheat plants, were sown at a density of four plants per 12 cm diam. pot. Plants were maintained in two blocks (1 pot/treatment/block) in a CE room at 20°C with a 16-hour photoperiod. After 1 week, the experimental plants were artificially inoculated by taking potted wheat plants infected with *Blumeria graminis* f. sp. *tritici* (formerly classified as *Erysiphe graminis* f. sp. *tritici*) (wheat PM), and trailing the infected leaves from these plants across the leaf surfaces of the healthy plants, such that spores from lesions on the infected leaves would brush off onto the healthy wheat plants. Treatment application (**Table 1**) commenced when the plants were 2 weeks old and at plant growth stage (PGS) 1, as defined by [37]. Leaves were sprayed to run-off (i.e., the point where the leaf is completely saturated and liquid starts to drip off the leaf) using a hand-held spray bottle (500 mL Garden Trigger Sprayer, Hills, Australia), with a total of 9 spray applications applied over a course of 7 weeks (2 sprays/week for the first fortnight, and 1 spray/week thereafter).

The first disease assessment (designated time 0) was made immediately before the first treatment application, followed by an assessment after 7 weeks (PGS = 8–10). Disease severity on the three most basal leaves of each plant was assessed using percent leaf area infection diagrams (**Figure 1**), and the rating scale shown in **Table 2**. Disease ratings for the three leaves were averaged to give one value per plant. Disease assessments had to be made on different leaves on each assessment date, because as the plants mature, the most basal leaves wither and die,

Treatment	Treatment code
Unsprayed control—no fungicides	Unsprayed
Water control	Water
Amistar® WG fungicide[1] (0.4 g/L)	Amistar
AMF[2] (7 g/L) + DATEM[3] (5 g/L) + Grindox[4] 122™ (1 g/L)	AMF
Soybean oil[5] (20 g/L) + DATEM (5 g/L) + Grindox 122™ (1 g/L)	SBO

[1]*Amistar® WG fungicide, containing 250 g/L azoxystrobin active ingredient, was supplied by Syngenta, Basel, Switzerland, and is effective against both powdery mildew and rust pathogens.*

[2]*AMF = anhydrous milk fat—a highly saturated solid milk fat, obtained from New Zealand Milk Products Ltd. (now trading as Fonterra).*

[3]*DATEM = an emulsifier containing diacetyl tartaric acid esters of mono- and di-glycerides. Sold by Danisco Ltd., Brabrand, Denmark as: Panodan® AL 10.*

[4]*Grindox 122™ = an antioxidant produced by Danisco Ltd., Brabrand, Denmark.*
[5]*Soybean oil (Amco brand) was obtained from the supermarket.*

Table 1.
CE room wheat trial treatments.

so data from each assessment date were analysed separately by SAS, version 8.02 (SAS Institute, Cary, NC), using a nested design, with treatments nested within pots and plants within treatments and pots.

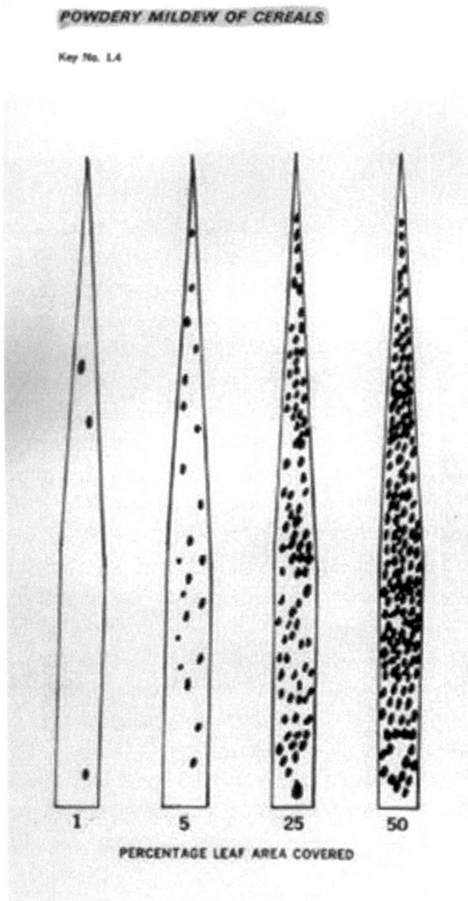

Figure 1.
Standard disease area diagrams to show four severities of wheat powdery mildew infection, from [37].

Rating	Percent leaf area infected
0	No infection
1	1% infection
1.5	1–5% infection
2	5% infection
2.5	5–25% infection
3	25% infection
3.5	25–50% infection
4	50% infection
4.5	>50% infection

Table 2.
Wheat powdery mildew (PM) leaf disease rating scale, from James [37].

2.2 CE trial results and discussion

At time 0, there were no consistent treatment differences evident (**Figure 2**), but after 7 weeks, "Amistar", "AMF" and "SBO" all provided significantly greater control of PM than "Unsprayed" and "Water" treated controls (**Figure 3**), and the amount of disease on plants treated with "AMF", "SBO" and "Amistar" was lower than that recorded at time 0, i.e. before any treatment application (**Figure 2** vs. **Figure 3**).

There were no visual signs of leaf damage associated with the treatments. (**Figure 4**).

Thus, under the controlled conditions of the CE room, SBO could perform as effectively as both the commercial fungicide Amistar and the AMF biofungicide,

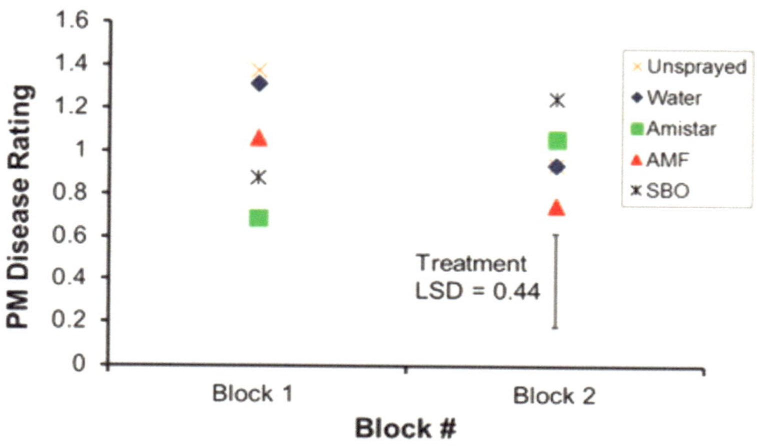

Figure 2.
PM disease severity on the basal leaves of 'Endeavour' wheat plants from the controlled environment (CE) wheat trial at time = 0, i.e. prior to any treatment application. Treatment codes are given in **Table 1***. The least significant difference (LSD) bar applies to within-column comparisons only, owing to the hierarchical nature of the nested design, where the number of replicate plants (n) for each data point on the graph = 4.*

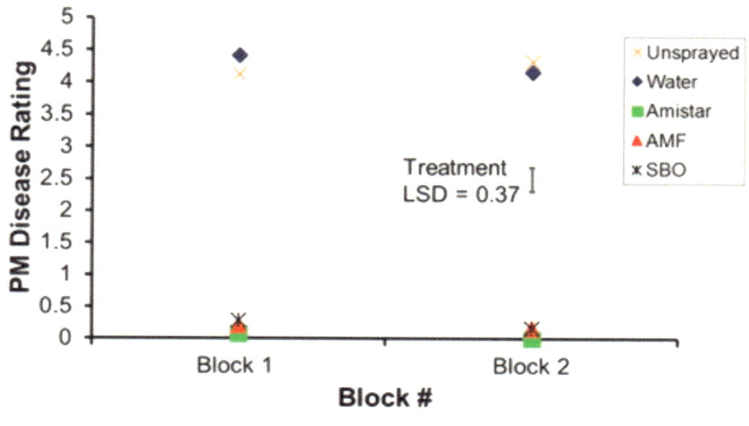

Figure 3.
PM disease severity on the basal leaves of 'Endeavour' wheat plants from the CE trial at time = 7 weeks, i.e. after 7 weeks of treatment application. Treatments are described fully in **Table 1** *and comprised leaving the plants unsprayed, or spraying with water, Amistar® fungicide or emulsified anhydrous milk fat (AMF) or soybean oil (SBO). The LSD bar applies to within-column comparisons only, owing to the hierarchical nature of the nested design, where n for each data point on the graph = 4.*

Figure 4.
'Endeavour' wheat plants in the CE room that have (A) not received any protection against PM (treatments "water" and "unsprayed"); or (B) were sprayed with "AMF", "SBO" or "Amistar" fungicides. The fungicides provided effective control of PM, without any visual adverse effects on plant health.

in terms of disease control and lack of adverse effects on plant health. Given that disease severity measured at the end of the experiment (after 7 weeks of treatment application) was lower than the disease severity measured at time 0 (before treatments commenced), this might suggest that all three fungicides have eradicant as well preventative activity against PM on wheat. However, this conclusion cannot be made for sure until assessments are made on whole plants throughout the course of the experiment, since different leaves were assessed at the start and the end of the CE experiment, owing to natural attrition of the oldest leaves.

2.3 Glasshouse trial methods

For the glasshouse trial (performed during February–March in Hamilton, New Zealand), the setup was similar to the CE trial, except that wheat seeds were sown into 6.75 L black polythene planter bags (PB12, Easy Grow Ltd., New Zealand). There were four replicate bags of four plants/treatment, and one replicate bag from each treatment was randomly positioned on a separate table (block). Treatments were the same as in the CE trial, except for omission of the unsprayed control. A total of seven spray applications were made throughout the course of the experiment (1 spray for the first fortnight, and 1 spray/week thereafter). Treatment application commenced when the plants were 11 days old and at plant growth stage (PGS) 1, as defined by [37].

Disease severity on the three most basal leaves/plant was assessed as described in the CE trial, with the initial disease assessment (designated Time 0) made immediately before the first treatment application, followed by an assessment after 7 weeks (PGS = 9–10.3). At the end of the trial, plants were considerably larger than those in the CE trial, so an additional disease assessment was also made at week 7 on the whole plant rather than the three most basal leaves, using the scale defined in [38], as shown in **Figure 5** and **Table 3**. Experimental design and statistical analysis was the same as for the CE trial.

2.4 Glasshouse trial results and discussion

Only data from the first (time = 0) and last (time = 7 weeks) disease assessments are presented. At time 0, disease levels in block 1 were significantly higher in the "Water" control than all other treatments, but this trend was not repeated in other blocks, and overall there were no consistent treatment differences evident at the start of the experiment (**Figure 6**). After 7 weeks, "Amistar", "AMF" and "SBO"

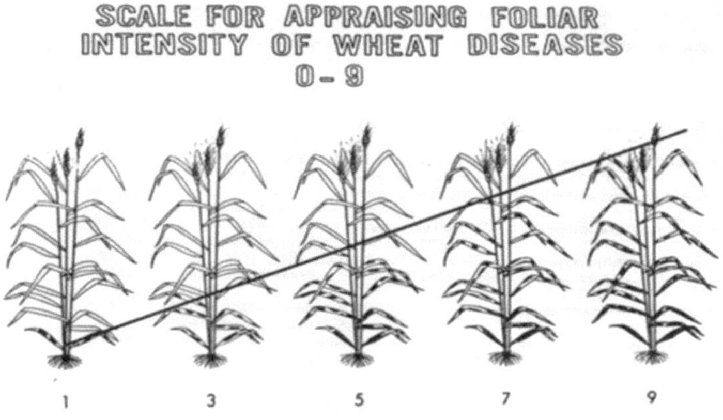

Figure 5.
PM disease severity on whole wheat plants, from [38].

Numeric scale	Characteristics
0	Free from infection.
1	Very resistant. Few isolated lesions on lowest leaves only.
2	Resistant. Scattered lesion on 2nd set of leaves, with first leaves infected at light intensity.
3	Moderately resistant. Light infection of lower third of plant.
4	Low intermediate. Moderate to severe infection of lower leaves, with scattered to light infection extending to the leaf immediately below the mid-point of the plant.
5	Intermediate. Severe infection of lower leaves, with moderate to light infection extending to the mid-point of the plant, but not beyond.
6	High intermediate. Severe infection of the lower third of the plant, moderate degree on middle leaves, and scattered lesions beyond the mid-point of the plant.
7	Moderately susceptible. Lesions severe on the lower and middle leaves, with infections extending to the leaf below the flag leaf, or with trace infections on the flag leaf.
8	Susceptible. Lesions severe on lower and middle leaves. Moderate to severe infection of upper third of plant. Flag leaf infected in amounts more than a trace.
9	Very susceptible. Severe infection on all leaves, and the spike infected to some degree.

Table 3.
Scale for appraising foliar intensity of wheat diseases on whole plants, from Saari & Prescott [38].

all provided significantly greater control of PM than "Water" treated controls, regardless of whether disease severity was measured on the three most basal leaves (**Figure** 7), or the whole plant (**Figure** 8). All three fungicides performed as well as each other. In all blocks, the amount of disease on whole plants treated with "AMF", "SBO" and "Amistar" was lower than that recorded at the start of the experiment, i.e. before any treatment application (**Figure** 7 c.f. **Figure** 8). This suggests that eradicant activity may be possible, under low initial inoculum loads (at the start of this experiment, there was <1% leaf infection) and corroborates the results found in the CE trial. However, the glasshouse environment is still relatively controlled and the plants are more widely spaced than in a field experiment, so field testing was the next step in the research.

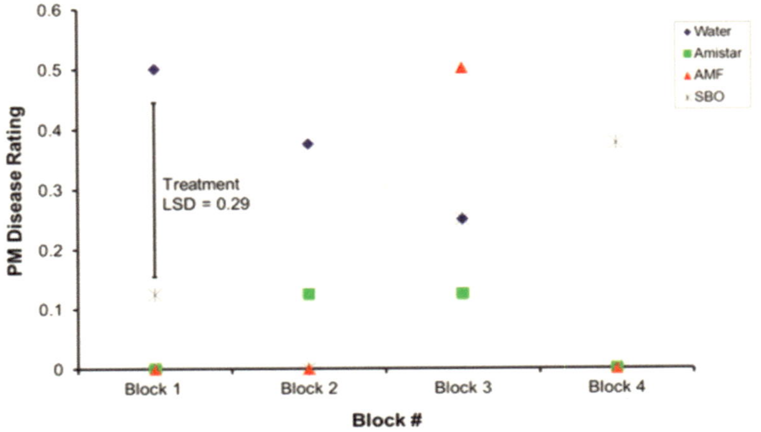

Figure 6.
Average PM disease severity on the basal leaves of 'Endeavour' wheat plants in the glasshouse trial at time = 0, i.e. prior to any treatment application. Treatment codes are given in **Table 1**. *The LSD bar applies to within-column comparisons only, owing to the hierarchical nature of the nested design, n for each data point on the graph = 4.*

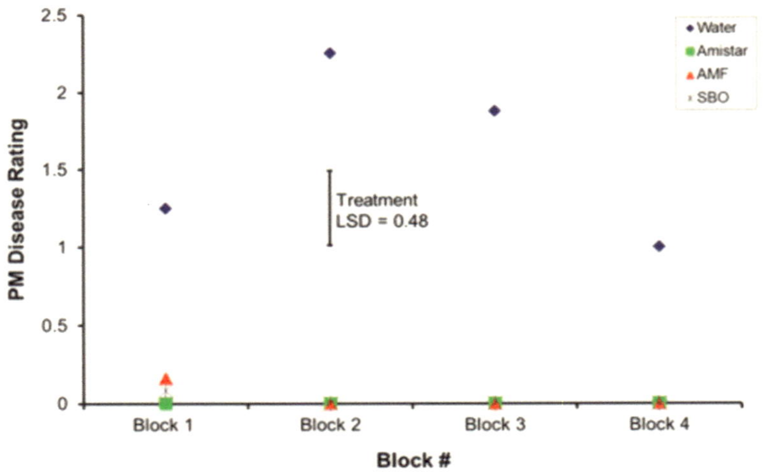

Figure 7.
Average PM disease severity on the basal leaves of 'Endeavour' wheat plants in the glasshouse trial at time = 7 weeks, i.e. after 7 weeks of treatment application. Treatment codes are given in **Table 1**. *The LSD bar applies to within-column comparisons only, owing to the hierarchical nature of the nested design, n for each data point on the graph = 4.*

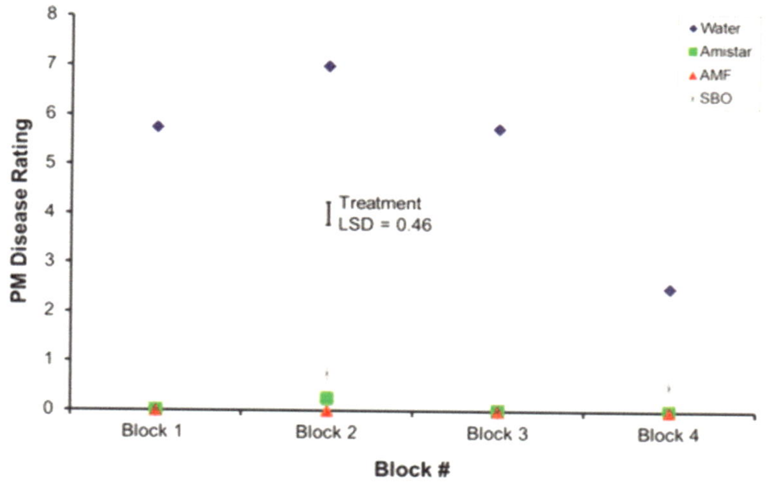

Figure 8.
Whole plant assessments of PM disease severity on 'Endeavour' wheat plants in the glasshouse trial at time = 7 weeks, i.e. after 7 weeks of treatment application. Treatment codes are given in **Table 1.** *The LSD bar applies to within-column comparisons only, owing to the hierarchical nested design, n for each data point on the graph = 4.*

3. Field trial

3.1 Field trial methods

Three spring wheat cultivars were used in this trial: 'Janz', an Australian cultivar highly susceptible to PM and resistant to brown leaf rust but susceptible to stripe rust; 'Karamu', a New Zealand cultivar that is susceptible to PM and leaf rust but resistant to stripe rust; and 'Gundaroi', a durum wheat cultivar that is susceptible to PM, but resistant to both rusts.

The three wheat cultivars were sown in separate adjacent areas in early September (spring) in Christchurch, New Zealand. Each cultivar was grown in two strips 24 m long and 1 m wide. Each strip was divided into 20 plots (1.2 m long × 1 m wide) and each plot contained approximately 150 plants. Adjacent, or nearly adjacent plots were randomly assigned to each of the five treatments to make a block and this procedure was repeated to give five blocks (25 plots), spread along the two strips, with block 3 split across both strips. The remaining 15 plots were untreated (buffers). Within each treatment plot five plants were labelled with block and treatment number. These labelled plants were used for repeat disease assessments over the trial period, with data analysed separately for each cultivar as a repeated measures design using SAS version 8.02 (SAS Institute, Cary, NC).

The five treatments were identical to those used in the CE trial, except that Amistar® fungicide was applied at the recommended field rate of 750 mL/ha and 1.4 mL of fungicide liquid concentrate /12 L water. During the growing season, there were five applications applied to designated treatment plots, sprayed to run-off using 20 L backpack pressure spray units (Backpack 435, Solo, New Zealand), of the water and biofungicide treatments (on average 17 days apart), and two applications of Amistar fungicide (7 weeks apart, according to manufacturer recommendations). After treatment application, plants were left to dry for several hours before disease assessments were carried out. At each assessment, the growth stage of each wheat cultivar was noted, as defined by James [37] . The five labelled plants in each plot were assessed according to a PM rating system from 0 to 9

(**Figure 5** and **Table 3**). The same rating system was used for a rust assessment on 'Janz', 123 days after sowing.

Plants were left in the field for 6½ weeks after the last spray until harvest in late February (summer). The trial was harvested with a rice binder (Model 210B, Mitsubishi, Japan) and each treatment/block rep was processed through a thresher (Nursery Master Stationary Thresher, Wintersteiger, Austria) to separate the wheat

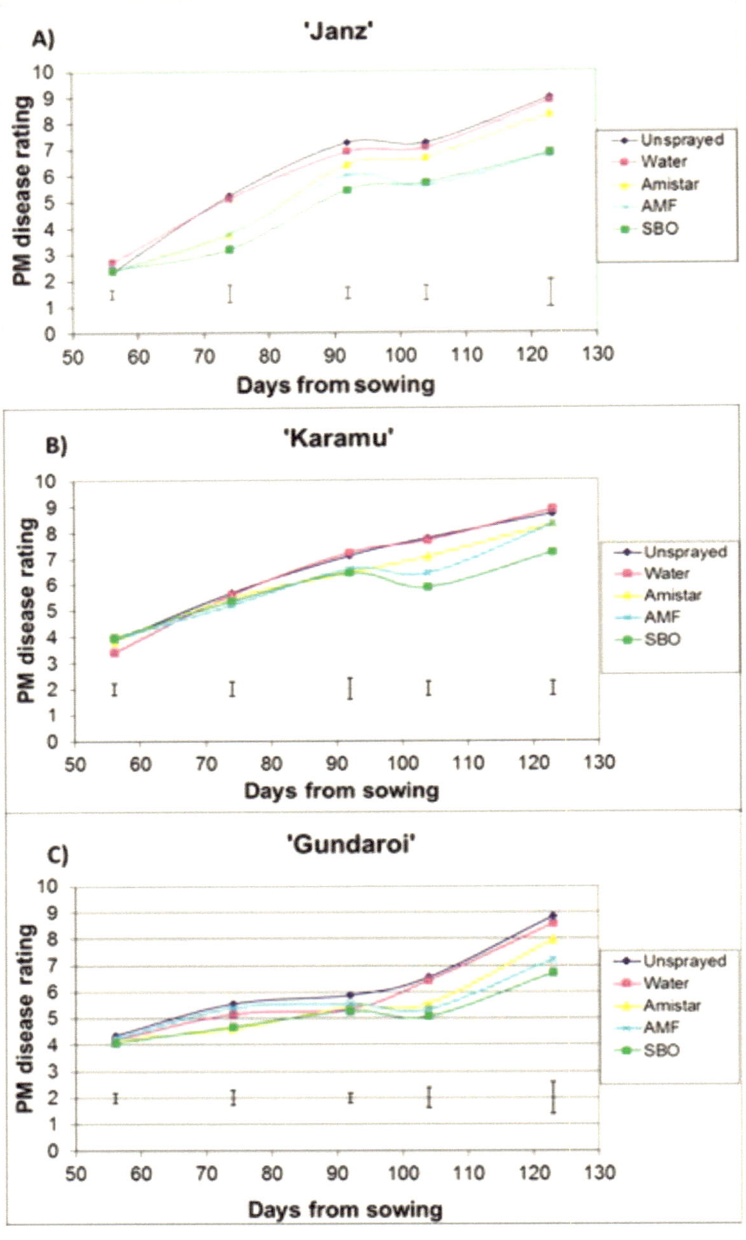

Figure 9.
Powdery mildew disease ratings in three field-grown wheat cultivars following application with Amistar® WG fungicide (750 mL product concentrate/ha at and 1.4 mL of fungicide liquid concentrate/12 L water), emulsified anhydrous milk fat (AMF) at 7 g/L and emulsified soybean oil (SBO) at 20 g/L. control treatments involved spraying the plants with water or leaving them unsprayed. Data were assessed as a repeated measures design with the bars indicating the least significance difference at each assessment date.

grain from the chaff and straw. The grain was placed into paper bags, which were taken back to the field lab and weighed. The following day, each bag was sorted through a 2 mm sieve screen (Endecott, United Kingdom) that separated grain into two lots: seconds (<2 mm) and first grade (>2 mm). The 1000 seed weights were measured on a machine (Numigral II, Tripette & Renaud, France) that automatically counts 250 seeds, which were weighed and then the weight was multiplied by four. Harvest data was analysed as a nested design (with blocks and treatments nested within cultivar) using SAS version 8.02.

3.2 Field trial results and discussion

For all three wheat cultivars, SBO was the most effective fungicide against PM, and provided significantly greater protection than the commercial synthetic fungicide, Amistar, during the middle part of the season, i.e. 90–120 days from sowing (**Figure 9**). The total degree of PM control was not as great as that observed in the CE and glasshouse trial, most likely because the close proximity of plants in the field trial resulted in overlapping growth leading to ineffective spray penetration and possible build-up of inoculum in protected parts of the canopy. The increase in disease was most marked in the final two disease assessments and under these heavy inoculum loads, Amistar was completely ineffective in the most PM-susceptible cultivar 'Janz' (**Figure 9**). No evidence of eradicant activity was observed for any of the products under the heavy inoculum loads and more variable conditions of the field trial.

Rust was only present in the 'Janz' cultivar, and SBO (and AMF) do not appear to provide control of this pathogen (**Figure 10**). Amistar claims to control rust, but there were no significant differences among the treatments (**Figure 10**).

Irrespective of wheat cultivar, SBO was associated with significantly lower harvest yields than all other treatments (**Table 4**). Yields in the AMF treatment were lower but not significantly different from the controls and plants sprayed with Amistar had significantly higher yields than both the other treatments (**Table 4**). This suggests that there is a yield cost associated with SBO and AMF use in the field trial. There are two possible explanations for this. The first is that oil use can be associated with damage to plant tissue, which affects the ability to produce and store photosynthates [13, 17–19]. However, we did not observe any chlorosis or necrosis associated with SBO or AMF use in any of our trials on wheat. More likely

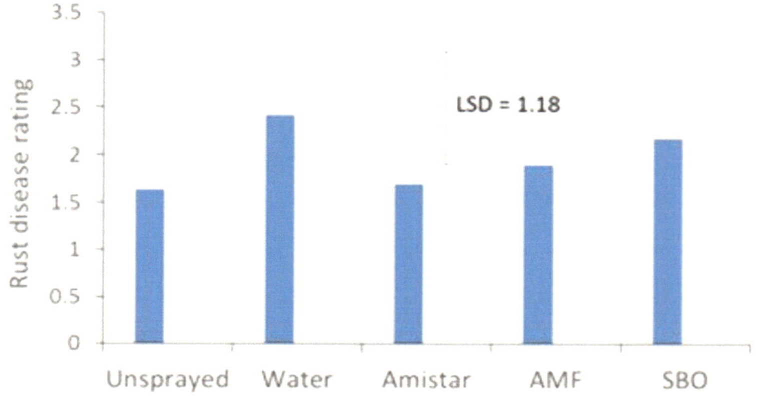

Figure 10.
Stripe rust levels in field-grown 'Janz' wheat, 123 days after sowing. Treatments are the same as described in **Figure 9***. Data were assessed as a randomised block design with means separation by least significant difference (LSD, P < 0.05).*

Cultivar	Treatment[1]	Total grain weight (g)	First grade weight[2] (g)	Weight 1000 grains (g)
'Janz'	Unsprayed	616	599	44.5
	Water	658	641	45.1
	Amistar	672	656	46.6
	AMF	592	575	42.6
	SBO	481	463	38.4
	LSD[3]	**69.4**	**67.2**	**2.15**
'Karamu'	Unsprayed	619	578	37.5
	Water	609	569	36.7
	Amistar	840	797	42.9
	AMF	605	570	37.2
	SBO	526	479	33.0
	LSD	**49.3**	**50**	**1.62**
'Gundaroi'	Unsprayed	619	578	37.5
	Water	609	569	36.7
	Amistar	840	797	42.9
	AMF	605	570	37.2
	SBO	526	479	33.0
	LSD	**49.3**	**50**	**1.62**

[1]*Control plants were left untreated or sprayed with water. Fungicide treated plants were sprayed with Amistar® fungicide (750 mL product concentrate/ha and 1.4 mL of fungicide liquid concentrate/12 L water), emulsified anhydrous milk fat (AMF) at 7 g/L and emulsified soybean oil (SBO) at 20 g/L.*
[2]*First grade grain has a size >2 mm.*
[3]*Least significant difference (P < 0.05).*

Table 4.
Yield data of field-grown wheat: Total grain weight (g), first grade grain weight (g) and weight of 1000 grains (g), harvested approximately 170 days after sowing.

is the second explanation that disease severity has escalated to a greater degree in the plants treated with lipid fungicide than in the Amistar treatment in the 6 1/2 week interval between the final spray application and harvest. Work in other crops (cucurbits and grapes) has shown that SBO and AMF need to be applied at fortnightly intervals to maintain effective disease control (Wurms, Plant & Food Research, unpublished data), whereas Amistar is a systemic fungicide (i.e., it is absorbed into the plant) and provides disease control over a more sustained period, and therefore only needed to be applied twice during the same trial period to provide effective control of PM [39] .

4. MoA studies

4.1 Scanning electron microscopy (SEM) methods

Experimental set-up was the same as for the CE trial, except that there was no water treatment, and there were two spray applications, 3 days apart. Four days after the second spray, lesions from all the treatments were sampled for electron microscopy.

Leaf pieces (5 × 10 mm) were cut from plants and mounted on a copper specimen stub, then processed for observation using a sputter cryo system (Emscope SP2000, Hemel Hempstead, United Kingdom). Mounted samples were first frozen using liquid nitrogen slush and then transferred under vacuum to a preparation chamber. There they were thermally etched for 5 min at −80°C, radiantly etched for 30–60 s, and then sputter coated with gold. The coated material was transferred under vacuum to a cold stage in the specimen chamber of a Philips SEM 505 scanning electron microscope (Philips, Eindhoven, Netherlands) and examined at an accelerating voltage of 15 kV and a specimen temperature of between −150 and −180°C [40].

4.2 SEM trial results and discussion

Amistar®, SBO and AMF fungicides all exhibited eradicant activity via a non-toxic (physical) MoA, as illustrated by direct effects on the fungus, since these treatments caused conidiophores (spore bearing structures) to collapse (**Figure 11**), conidia (asexual spores) to wither (**Figure 12**) and extrude cellular contents (**Figures 12** and **13**) and hyphae to wither/desiccate (**Figure 13**). This is supported by the CE and glasshouse trial data which showed that, under low initial inoculum loads, disease severity decreased on the same wheat plants monitored over time (**Figures 2, 3, 6–8**). The MoA of SBO is most likely created by disruption of membrane transport of the pathogen since the SEM images indicate that SBO causes plasmolysis of mycelia and cell rupture and leakage of cell contents, especially in conidia

Figure 11.
SEM images of powdery mildew colonies on 'Endeavour' wheat leaves that were either A, untreated; or sprayed with: B, Amistar® fungicide; C, anhydrous milk fat (AMF); or D, soybean oil (SBO). In healthy, unsprayed colonies (A), turgid hyphal threads can be seen growing along the leaf surface in among pointy/tapered trichomes (leaf hairs), an example of which is arrowed in A, and upright conidiophores bearing chains of spherical conidia (asexual spores) are visible extending upwards and outwards from the leaf surface. Conversely, hyphae appear shrivelled and conidiophores have completely collapsed in sprayed colonies (B-D).

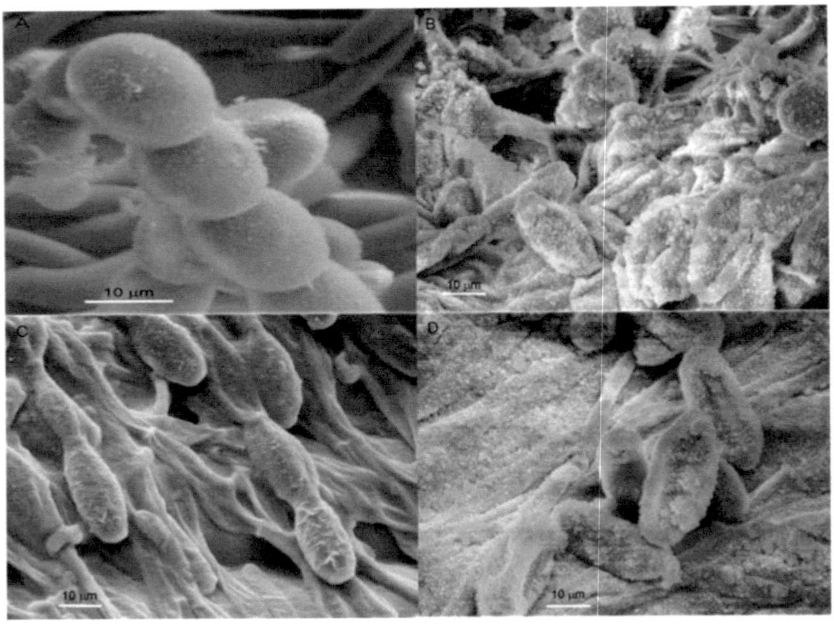

Figure 12.
Higher magnification SEM images of powdery mildew conidia on 'Endeavour' wheat leaves that were either A, untreated; or sprayed with: B, Amistar® fungicide; C, AMF; or D, SBO. In unsprayed plants (A), the conidia are present in chains attached to conidiophores that protrude outwards from the leaf surface. Unsprayed conidia have a plump/turgid appearance and the spore surface appears to be quite smooth. In contrast, conidia on sprayed plants are lying collapsed on the leaf surface and have a withered/dehydrated appearance (B-D). Ridging of the conidial surface is apparent in AMF-treated plants (C), and grainy exudates, most probably cell contents, surround conidia sprayed with fungicide (B) and soybean oil (D).

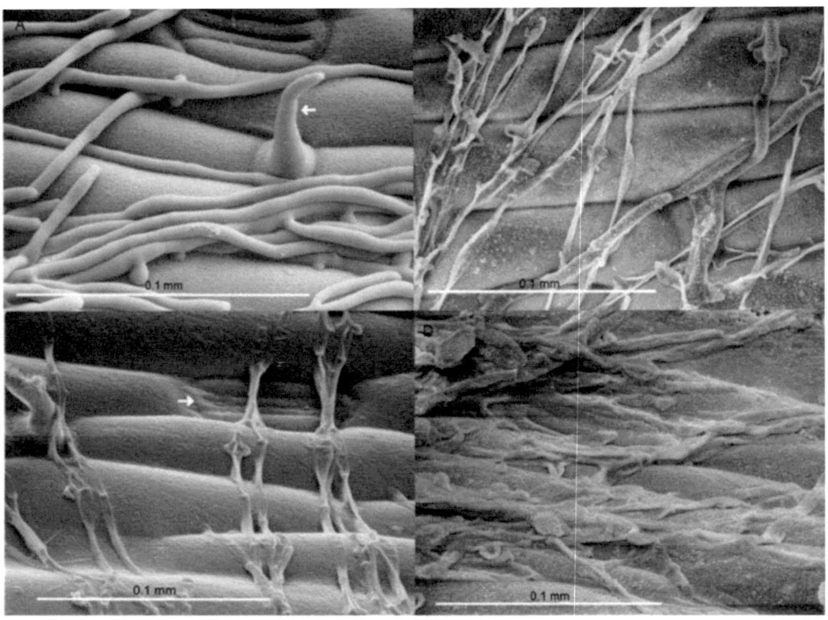

Figure 13.
SEM close-ups of powdery mildew hyphae on 'Endeavour' wheat leaves that were either A, untreated; or sprayed with: B, Amistar® fungicide; C, AMF; or D, SBO. In unsprayed plants (A), the hyphae growing on the leaf surface are plump and turgid. A trichome is arrowed in A. By contrast, hyphae on sprayed plants are completely withered/dehydrated (B-D). A plant stomate (pore for gaseous exchange) on the leaf surface is arrowed in C.

(**Figures 11–13**). This is in agreement with a review on the antimicrobial mode of action of essential plant oils, where antimicrobial action was most commonly due to membrane permeabilization/disruption leading to loss of water, leakage of cell contents and sometimes total lysis [41] . In contrast, AMF may have a different MoA to SBO, as SEM demonstrated that AMF caused deep ridging and distortion of conidia, rather than extrusion of cell contents (**Figures 11–13**). Combination of products with different complimentary MoA has the potential to increase product durability and efficacy, since it is more difficult for the pathogen to develop resistance. In addition, pathogens do not tend to develop resistance to agricultural sprays containing oils [30, 31, 42], because membrane transport is such a fundamental life process. The combination of reduced amounts of SBO and AMF has been shown to be as effective as greater concentrations of each biofungicide on its own [10]. Non-toxic (physical) rather than toxic (chemical/antibiotic) MoAs are also advantageous when it comes to product registration, since the latter higher-risk product group requires expensive and time consuming toxicology testing.

5. Commercialisation potential

SBO biofungicide shows great commercial promise, given that it provided PM control on wheat in both controlled internal conditions and more variable field environments (at least for part of the growing period) that equalled or exceeded that provided by conventional fungicides. Moreover, SBO is cost effective and simple to produce (data not presented), has a physical mode of action (thus making registration easier) and contains Generally Recognised As Safe (GRAS) [43], edible ingredients. However, there are critical factors that must be taken into consideration to ensure optimal performance and success of this product.

The most important consideration for use of this type of fungicide is that it has a direct, non-systemic mode of action [44] and therefore requires direct contact with the plant surface that it is to protect. This explains why disease control was much higher in the CE and glasshouse trials, where plants were more widely spaced, and why PM pustules were observed on the undersides of leaf blades close to the stems of plants in the field trial (data not presented), because heavily overlapping foliage prevented spray access. Our research has also shown that the percentage of grape bunch surfaces directly exposed to spray (i.e., not covered by leaves), as determined by leaf plucking, was a significant factor in the efficacy of SBO fungicide against *Botrytis cinerea* [11]. Consequently spray penetration and the density of plant architecture/growing systems are key considerations to the success of this fungicide.

We believe that other disadvantages that may be associated with SBO can be managed with careful use. Although phytotoxicity is sometimes associated with oils, optimisation of formulation (as well as rate and frequency of application) has been shown to minimise toxic effects [11, 45]. Our SBO formulation [46] has managed to achieve the balance of efficacy without adverse effects on plant health. Phytotoxicity can also be avoided by taking care not to tank mix products such as elicitors [32] or sulphur [44, 47] as these may react together to form plant damaging compounds causing foliar injury and leaf drop. However, these products can still be successfully used together in an integrated spray programme provided that their use is alternated [11]. SBO has also been demonstrated to have a much less adverse effect on plant health than AMF [32]. Another effective option is that SBO can be tank mixed with AMF at much lower concentrations than either product on its own [10]. This offers the dual advantages of reduced cost of goods and greater durability, due to differing modes of action as described in the preceding SEM section. Other recommendations include not spraying below 4°C (40°F), because the emulsion breaks down, and avoiding sprays on newly emerged foliage or floral tissue, although we have treated rose blooms without any toxic effects [10].

Although SBO exhibited both preventative and eradicant activity in this study, eradicant activity was not effective against heavy, established inoculum loads in the field trial, and consequently SBO is best used as a preventative. Given that horticultural oils degrade readily and are not very persistent [31], they also need to be applied at regular (e.g. fortnightly) intervals. A lack of PM control over the last 6 weeks between spraying and harvest could be the reason for loss of yield in the wheat field trial, although further work would need to be carried out to confirm this by carrying out a PM disease assessment at harvest. SBO is particularly well-suited for use in an integrated pest management (IPM) programme. In New Zealand, our SBO formulation has been registered as MIDI-Zen® by BotryZen 2010 Limited, and is intended to be used a part of a residue-free IPM programme for control of *Botrytis cinerea* on grapes. Although use of MIDI-Zen right up to vintage in grapes has been shown to delay the increase in soluble sugars (°Brix), which would necessitate delaying harvest for 1–2 weeks to allow Brix to rise, this problem is normally avoided by using MIDI-Zen in the middle part of the grape growing season (from pre-bunch closure to version) and another biofungicide in the final 3 weeks leading up to harvest [11].

In summary, the potential for SBO is very exciting as it offers the potential for effective, environmentally benign and durable control. Armed with a good formulation and an understanding of how best to optimise its use and minimise any adverse effects, we should see increased use of SBO in agriculture to improve plant health.

Acknowledgements

Special thanks to Nicole Shukker, former employee of NZ Milk Products Ltd. (now Fonterra), to Lisa Marcroft, former business manager (BM) at HortResearch, and to Claire Hall (current BM at Plant & Food Research Ltd) for their drive and enthusiasm for this project. We are very grateful to Dr. Siva Ganesh for guidance with statistics and use of SAS software, and to Mike Spiers and Dr. Philip Elmer for reviewing this manuscript. This research was funded by the Pre-Seed Accelerator Fund, New Zealand.

Author details

Kirstin V. Wurms[1]*, Annette Ah Chee[1] and Paul Sutherland[2]

1 The New Zealand Institute for Plant and Food Research Limited, Hamilton, New Zealand

2 The New Zealand Institute for Plant and Food Research Limited, Auckland, New Zealand

*Address all correspondence to: kirstin.wurms@plantandfood.co.nz

References

[1] Encyclopaedia Britannica. Soybean plant. [Internet]. 2018. Available from: https://www.britannica.com/plant/soybean [Accessed: 2018-06-01]

[2] Organic Facts. Thirteen amazing benefits of soybeans. [Internet]. 2018. Available from: https://www.organicfacts.net/health-benefits/cereal/soybeans.html [Accessed: 2018-06-01]

[3] Johnson LA. 11-Oil recovery from soybeans. In: Johnson LA, White PJ, Galloway R, editors. Soybeans. Chemistry, Production, Processing and Utilisation. Urbana: AOCS Press; 2008. pp. 331-375. DOI: 10.1016/B978-1-893997-64-6.50014-7

[4] Wikipedia. Soybean oil. [Internet]. 2018. Available from: https://en.wikipedia.org/wiki/Soybean_oil [Accessed: 2018-06-01]

[5] Soil Service Inc. Landoil. [Internet]. 2018. Available from: http://soilserviceinc.com/spray-surfactants/landoil/ [Accessed: 2018-06-06]

[6] Somervaille A, Betts G, Gordon B, Green V, Burgis M, Henderson R. Adjuvants—Oils, surfactants and other additives for farm chemicals. [Internet]. 2012. Available from: https://grdc.com.au/__data/assets/pdf_file/0017/222155/adjuvants-oils-surfactants-and-other-additives-for-farm-chemicals-revised-2012-edition.pdf.pdf [Accessed: 2018-06-06]

[7] Ko WH, Wang SY, Hsieh TF, Ann PJ. Effects of sunflower oil on tomato powdery mildew caused by *Oidium neolycopersici*. Journal of Phytopathology-Phytopathologische Zeitschrift. 2003;**151**:144-148. DOI: 10.1046/j.1439-0434.2003.00698.x

[8] Northover J, Schneider KE. Activity of plant oils on diseases caused by *podosphaera-leucotricha*, *venturia-inaequalis*, and *albugo-occidentalis*. Plant Disease. 1993;**77**:152-157. DOI: 10.1094/pd-77-0152

[9] Wurms KV, Ah Chee A. Control of powdery mildew (*Podosphaera leucotricha*) on apple seedlings using anhydrous milk fat and soybean oil emulsions. New Zealand Plant Protection. 2011;**64**:201-208

[10] Wurms KV, Hofland-Zijlstra JD. Control of powdery mildew on glasshouse-grown roses and tomatoes in the Netherlands using anhydrous milk fat and soybean oil emulsions. New Zealand Plant Protection. 2015;**68**:380-388

[11] Wurms K, Ah Chee A, Elmer P, Agnew R, Wood P. Developing new biologically-based products for control of botrytis bunch rot. Part 1: Developing a new natural product for mid-season botrytis control—NP2 moves closer to the market. Wine and Viticulture Journal. 2011;**26**:64-72

[12] Xuan TD, Gu GQ, Minh TN, Quy TN, Khanh TD. An overview of chemical profiles, antioxidant and antimicrobial activities of commercial vegetable edible oils marketed in Japan. Food. 2018;**7**(2):21. DOI: 10.3390/foods7020021

[13] Moran RE, Deyton DE, Sams CE, Pless CD, Cummins JC. Soybean oil as a summer spray for apple: European red mite control, net CO_2 assimilation and phytotoxicity. Hortscience. 2003;**38**:234-238

[14] Butler GD, Coudriet DL, Henneberry TJ. Toxicity and repellency of soybean and cottonseed oils to the sweetpotato whitefly (homoptera, aleyrodidae) and the cotton aphid (homoptera, aphididae) on cotton in greenhouse

studies. Southwestern Entomologist. 1988;**13**:81-86

[15] Fenigstein A, Eliyahu M, Gan-Mor S, Veierov D. Effects of five vegetable oils on the sweetpotato whitefly *Bemisia tabaci*. Phytoparasitica. 2001;**29**:197-206. DOI: 10.1007/bf02983451

[16] Shaaya E, Kostjukovski M, Eilberg J, Sukprakarn C. Plant oils as fumigants and contact insecticides for the control of stored-product insects. Journal of Stored Products Research. 1997;**33**:7-15. DOI: 10.1016/s0022-474x(96)00032-x

[17] Baysal-Gurel F, Miller SA. Management of powdery mildew in greenhouse tomato production with biorational products and fungicides. In: Paret ML, Vallad GE, Zhang S, Jones JB, editors. IV International Symposium on Tomato Diseases; Acta Horticulturae. Vol. 1069. 2015. pp. 179-183

[18] Finger SA, Wolf TK, Baudoin AB. Effects of horticultural oils on the photosynthesis, fruit maturity, and crop yield of winegrapes. American Journal of Enology and Viticulture. 2002;**53**:116-124

[19] Zhuang QG, Wang LH, Li MZ, Hou TP, Xie Y. Enspray 99′ mineral oils for white peach scale, *Pseudaulacaspis pentagona* and phytotoxicity to 'Hongyang' kiwifruit. In: Huang H, Zhang Q, editors. VIII International Symposium on Kiwifruit; Acta Horticulturae. Vol. 1096. 2015. pp. 363-369

[20] Glawe DA. The powdery mildews: A review of the world's most familiar (yet poorly known) plant pathogens. Annual Review of Phytopathology: Annual Reviews. 2008;**46**:27-51. DOI: 10.1146/annurev.phyto.46.081407.104740

[21] Pasini C, Daquila F, Curir P, Gullino ML. Effectiveness of antifungal compounds against rose powdery mildew (*Sphaerotheca pannosa* var. *rosae*) in glasshouses. Crop Protection. 1997;**16**:251-256. DOI: 10.1016/s0261-2194(96)00095-6

[22] Grains Research & Development Corporation. Barley powdery mildew. Fact sheet [Internet]. 2012. Available from: https://grdc.com.au/__data/assets/pdf_file/0028/24958/grdc-fs-barley-powdery-mildew.pdf [Accessed: 2018-06-06]

[23] Turechek WW, Mahaffee WF, Ocamb CM. Development of management strategies for hop powdery mildew in the Pacific Northwest. Plant Health Progress. 2001;**2**:1. DOI: 10.1094/PHP-2001-0313-01-RS

[24] Fuller KB, Alston JM, Sambucci OS. The value of powdery mildew resistance in grapes: Evidence from California. Wine Economics and Policy. 2014;**3**:90-107. DOI: 10.1016/j.wep.2014.09.001

[25] Colcol JF, Rallos LE, Baudoin AB. Sensitivity of *Erysiphe necator* to demethylation inhibitor fungicides in Virginia. Plant Disease. 2012;**96**:111-116. DOI: 10.1094/pdis-12-10-0883

[26] Lopez-Ruiz FJ, Perez-Garcia A, Fernandez-Ortuno D, Romero D, Garcia E, de Vicente A, et al. Sensitivities to DMI fungicides in populations of *Podosphaera fusca* in south Central Spain. Pest Management Science. 2010;**66**:801-808. DOI: 10.1002/ps.1948

[27] Ye M, Beach J, Martin JW, Senthilselvan A. Occupational pesticide exposures and respiratory health. International Journal of Environmental Research and Public Health. 2013;**10**:6442-6471. DOI: 10.3390/ijerph10126442

[28] Homestead Organics. Sulphur. [Internet]. 2018. Available from: http://www.homesteadorganics.ca/sulphur_pest.htm [Accessed: 2018-06-07]

[29] Brunetto G, Ferreira PAA, Melo GW, Ceretta CA, Toselli M. Heavy metals in vineyards and orchard soils. Revista Brasileira De Fruticultura. 2017;**39**:2(e-263). DOI: 10.1590/0100-29452017263

[30] Horticentre Group. Eco-oil. [Internet]. 2018. Available from: http://www.horticentre.co.nz/page/23-products-services+pest-disease-control+crop-chemistry-biologicals-growth-aids-and-cleaners+insecticides+insecticides-d-f+eco-oil [Accessed: 2018-06-07]

[31] Skinner A. Spraying with horticultural oils. [Internet]. 2017. Available from: http://ucanr.edu/datastoreFiles/268-754.pdf [Accessed: 2018-06-01]

[32] Ah Chee A, George M, Alavi M, Wurms K. Lipid based fungicides for control of powdery mildew in cucurbits. New Zealand Plant Protection. 2018;**71**:262-271. DOI: 10.30843/nzpp.2018.71.123

[33] Australian Wine Research Institute. Understanding chemical 'modes of action' [Internet]. 2010. Available from: https://www.awri.com.au/wp-content/uploads/spray_modes_of_action.pdf [Accessed: 2018-09-10]

[34] McGrath M. Fungicides and mode of action [Internet]. 2005. Available from: https://gpnmag.com/article/fungicides-and-mode-action/ [Accessed: 2018-09-10]

[35] AHDB Cereals & Oilseeds. Powdery mildew [Internet]. 2018. Available from: https://cereals.ahdb.org.uk/cereal-disease-encyclopedia/diseases/powdery-mildew.aspx [Accessed: 2018-06-20]

[36] Salgado JD, Paul PA. Powdery Mildew of Wheat [Internet]. 2016. Available from: https://ohioline.osu.edu/factsheet/plpath-cer-11 [Accessed: 2018-06-21]

[37] James WC. An illustrated series of assessment keys for plant diseases, their preparation and usage. Canadian Plant Disease Survey. 1971;**51**:39-65

[38] Saari EE, Prescott JM. Scale for appraising foliar intensity of wheat diseases. Plant Disease Reporter. 1975;**59**:377-380

[39] Syngenta. Amistar [Internet]. 2018. Available from: https://www.syngenta.co.nz/product/crop-protection/fungicide/amistar [Accessed: 2018-09-13]

[40] Falloon RE, Sutherland PW, Hallett IC. Morphology of *Erysiphe pisi* on leaves of *Pisum sativum*. Canadian Journal of Botany. 1989;**67**:3410-3416

[41] Hyldgaard M, Mygind T, Meyer RL. Essential oils in food preservation: Mode of action, synergies, and interactions with food matrix components. Frontiers in Microbiology. 2012;**3**:24. DOI: 10.3389/fmicb.2012.00012

[42] Halcomb M, Hale F. General Comments about the Use of Horticultural Oil [Internet]. 2012. Available from: https://extension.tennessee.edu/mtnpi/Documents/handouts/Insect%20and%20Disease%20Control/Horticultural_Oil_Handout.pdf [Accessed: 2018-06-21]

[43] U.S. Food & Drug Administration. Generally recognised as safe (GRAS) [Internet]. 2018. Available from: https://www.fda.gov/food/ingredientspackaginglabeling/gras/ [Accessed: 2018-09-13]

[44] Pundt L. Integrated pest management program. Horticultural oils. [Internet]. 2015. Available from: http://ipm.uconn.edu/documents/raw2/html/831.php?aid=831 [Accessed: 2018-06-21]

[45] Wurms K, Ah Chee A. Product formulation is crucial to the success

of lipid-based bio-fungicides. New Zealand Plant Protection. 2018;**71**:272-284. DOI: 10.30843/nzpp.2018.71.124

[46] Wurms KV, Ah Chee A. Fungicidal composition comprising anhydrous milk fat (AMF) and soybean oil for the treatment of powdery mildew. Horticulture & Food Res Inst assignee. NZ patent NZ534007A; 2007

[47] Extension Toxicology Network. Sulfur. [Internet]. 1995. Available from: http://pmep.cce.cornell.edu/profiles/extoxnet/pyrethrins-ziram/sulfur-ext.html [Accessed: 2018-06-21]

Chapter 2

Yield Components and Biomass Partition in Soybean: Climate Change Vision

Milton E. Pereira-Flores and Flávio B. Justino

Abstract

Long-term climate change and inter-annual climate variability are events of concern to farmers and humanity. Global warming could affect agriculture in various ways and it is anticipated that agricultural systems will face great pressure from the variability of climate factors and their extreme events, which in most cases are difficult to predict, particularly extreme events of rainfall, higher dry season, hot and cold waves and their interactions. Global warming could also have some positive effects for plants such as increasing the temperature of current cold regions and increasing carbon dioxide with its positive effect on photosynthesis, growth rates, the use of water and production. Meanwhile, there are still many questions that remain about this possible future. This chapter, brings the response of plants to future conditions through specifics alterations in its components of yield on environmental conditions with enrichment of CO_2 and elevated temperature, two climatic factors, which is understood to be the factors of climatic change of greater global extent. The study of the components of yield and their alterations, can guide diverse sectors of the sciences and decision makers, in order to structure strategies of resilience in the cultivation of soybean.

Keywords: yield components, soybean, global climate change, elevated CO_2 and temperatures, production

1. Introduction

Global climate models predict increases in air temperature by up to 2–4°C, CO_2 concentrations higher than 700 mmol.mol^{-1} and increase in ground O_3 higher than 70 ppb by the end of the year 2100 [1, 2].

Based on these projections about the changes in the growth environment of the cultivated plants, it will be prudent to know how the current cultivars can be affected in their yield components, which are what define the productive potential.

Despite the existence of many studies simulating future scenarios made in FACE (Free-CO_2 environmental), OTC (open top chamber) and Growth Chambers to know how the altered climate factors will affect the physiology and production of soybean, few studies have been directed to understand those alterations in the level of the yield components that are the intrinsic factors of the plant more sensitive to climate change, and that also depend on the management of the crop at the field level. This anticipated knowledge may be important for the direction of policies and research lines in various areas of agricultural sciences to develop diverse resilience strategies to climate change.

The understanding of how climate influences the growth, development and production of soybean plants depends on the understanding of how the yield components respond to the variations of climate factors, which can also be elucidated if studies the plant alterations in the future atmosphere conditions. The plant production is determined by changes in yield components, in last instance. The artificial enrichment of the growth environment of soybean plants with CO_2, O_3 and temperatures according to the forecasts on the atmospheric composition for the year 2100, can allow to know the morphophysiological responses in several levels of the plant organization, long before environmental changes occur.

The study of the morphophysiological mechanisms of response of soybean plants to the ecological environment where they develop and produce grains, constitutes the basis of soybean ecophysiology.

The factors of the climate (temperature, radiation, rainfall, wind and atmospheric pressure, among others), plus the physicochemical properties of the soil and the cultural practices applied in the field continuously influence the performance of the community of soya plants from germination to the senescence of the plants. Throughout the different phenological stages, the expression of stage-tissue genes defines the course of the development of the plant, the formation of the biomass and its components (roots, stems, leaves, flowers, fruits and seeds) respond simultaneously and hierarchically with the objective of completing its biological cycle and producing seeds for the perpetuation of the plant, and that for humanity represents the basis of agricultural production.

The most important climatic factors for the development and production of soybean cultivars are temperature and photoperiod and their interactions, plus other favorable/limiting factors and resources such as precipitation variability, appropriate supply of nutrients and elimination of inter- and intraspecific competition, which also interact to determine the production of soybeans in a given region [3, 4].

The temperature is directly related to the speed of the metabolic rates and the chronological duration of the different phenological stages of the crop, and in the case of floral induction, in interaction with the photoperiod in plants responsive to the duration of the night to flower. The photothermal influence in the growth stages can be predicted by unit heats method. In general, the temperature determines the growth rates and the duration, in days, of each stage of the development phases. Soybeans have cardinal temperatures for most of their developmental stages [5, 6].

According to the American Society of Meteorology, cardinal temperatures correspond to the minimum (Tb) and maximum (TB) temperatures that define the limits of growth and development of an organism and an optimal temperature (Top) in which growth proceeds more rapidly (htpp://glossary.ametsoc.org/wiki/Cardinal_temperatures). According to the above, the rate of development increases linearly between Tb and Top; decreases from Top to TB, and after TB the development stops and the duration of the phase becomes infinite. It is possible to specify that Top, is not a thermal point, but the average of a very narrow range of temperatures, where the majority of the enzymatic reactions that participate in the growth is close to their catalytic maximums.

The soybean cardinal temperatures defined several plant processes from temperature thresholds. The lower base temperature (Tb) vary between 6 and 10°C to plant development. The lower thermal thresholds are: (Tb) of 10°C [7], 11°C [8], and 14°C [9].

The germination rate is close to zero at 15°C and maximum at 25°C [10]. Other thermal threshold to highest growth was found between 29 and 31°C [11, 12], thermal threshold that is the same to maxima protein content in the grains [13].

For photosynthesis, optimum diurnal temperatures are between 30 and 35°C, and for growth, night temperatures between 21 and 27°C. Temperatures less than 22°C delay the retention of pods and at temperatures ≤14°C flower abortion may occur [14]. Temperatures close to or above 40°C have negative effects on growth rate and pod retention [15].

In general, vegetative and reproductive growth in soybean can reach high rates in temperatures between 25°C and 30°C during the growth season, because the maximal vegetative and reproductive development occur in 30°C and 25-29°C, respectively. In addition the optimal floral anthesis temperature is achieved in 26°C [16, 17]. Thus, the choice of the time of year considering the regime of soil and air temperatures are determinant to establish the best sowing times, in which the thermal supply of the soil-atmosphere system is satisfactory, together with the adequate availability of water to meet the consumptive use of the crop. In fact, soybean yield components are negatively correlated with temperature increase and these components are temperature-dependent [18, 19], mainly, when the temperatures on field exceed the optimum temperatures [20, 21].

The main effect of the photoperiod is to induce flowering after the juvenile phase is over. Low temperatures and long days delayed the flowering time, and consequently, the anthesis and the maturation time [22, 23]. This relation is widely known, as well as, that the greater sensitivity to photoperiod and low temperatures are more obvious among the genotypes with greater sensitivity to the photoperiod; late maturation cultivars are more sensitive than early cultivars [24]. Most of the soybean cultivars have a pre-inductive or short juvenile stage, and floral induction may occur at any stage after the development of the first unifoliate leaves [25]. With the incorporation of long juvenile periods, soybeans currently produce soybeans until the 15th degree of Latitude, preventing the early induction of flowering [26].

The variation of flowering time between soybean cultivars, from a genetic point of view is very complex, because it will probably not be so easy to identify the molecular bases of the major genes and Quantitative Trait Loci (QTLs) underlying the natural variation in flowering time of soybean, because most of those genes and QTLs exist in multiple copies in the genome, interacting more or less with one another and with the environments in which the genes are evaluated [27]. In the specific case of soybean, some cultivars must fulfill a juvenile stage before the influence of the photoperiod for the induction of flowering, and the sensitivity can occur from the expansion of the first V1-V2 vegetative stages (first and second trifoliate leaf) [28]. From then on, the taxa of growth and development of the plants will be a function of the availability of light, water, nutrients, and above all, of the temperature up to values close to the optimal Day/Night temperature. In turn, after flowering induction plus higher temperatures the duration of this inductive stage can be varied and influence the size and characteristics of the canopy, that is, the height of plants, the number and length of productive branches, effective leaf area and number of flowers per cluster, among others.

Rainfall is the most common form of water supply, and its intensity and variability are pointed out as determinants of the risk to the success of production in most soy producing regions in Brazil, EUA and China [29]. Precipitations between 450 and 800 mm may allow high yields depending on distribution throughout the cultivar cycle and on edaphic and climatic conditions [30]. However, this high yield potential is soil type and climate dependent, mainly of the interaction with temperature, evapotranspiration, and soil water content. The inter-annual variability of those climatic factors provoked by the climatic changes are characterized by the occurrence of extreme events of excess and precipitation deficit and heat waves in relation to the normal climatological is great determinant of the soybean yield [31, 32]. The occurrence of prolonged "veranicos" (absence of rainfall for 25 continuous days or more during the summer) has been more frequent and prolonged, for example, in December and February in the central region of Brazil and pointed as the most dangerous condition for the success of soybean plantations.

2. Yield and yield components

2.1 Yield production

The increase in soybean production under high [CO_2] has been variable, ranging from increases of about 17%, marginal increases [33, 34], and no gain in production [35]. In most cases the increases have been derived mainly from gains in the total weight of grains at harvest, the increase in the number of pods [36] and the average weight of the grains [34, 37].

In understanding the magnitude of differences in production gains over high CO_2 concentrations, we should consider aspects such as the type of cultivar, production system and densities used, and the interaction with the climatic factors of each region where the plantations occurred. For example, it was verified increases in biomass and seed weight in the day/night thermal regimes 20/15°C compared with elevated thermal day/night regime like 30/25°C under 700 ppm CO_2 [38]. In other similar study, a greater number of branches and productive nodes were formed in 26/20 than in 22/16°C [39]. In this case, the positive interaction between elevated [CO_2] and temperatures regimes, resulting in increases in production. Thus, it can be concluded that the closer to the temperature regime of the optimum temperature, positive interaction can be expected for greater production, than when the temperatures exceed the Top. However, the meta-analysis performed on the results of several studies on the productive response of soybeans to CO_2 increase shows that, despite increases in foliar absorption of CO_2, soybean production is less responsive in experimental conditions and that the responses in field conditions were smaller than those performed in confinement (pot use) [37] The question, again, goes back to the point of knowing how to explain this low response at plant level.

Recently, a study conducted with 18 soybean cultivars (II, III, IV soybean groups) conducted in several years repeated with 550 ppm of CO_2, found average responses of 22% increase in the aerial biomass and only 9% in the yield of the seed, when grown in the appropriate growing season, and average temperatures of the growing season varying between 20.7 and 23.3°C [40]. During 4 years of study, there was consistency from year to year among genotypes that were more and less sensitive to the elevation of [CO_2], suggesting heritability of the CO_2 response [40]. In addition, cultivars with the highest coefficient of partition to the seed in the current [CO_2] also had the highest partition coefficient in the high [CO_2] [40]. This suggests, the existence of a variation genetic in the response of soybean to a high level of [CO_2], which is necessary to obtaining cultivars of soybean that adapt to future conditions.

2.2 Yield component basis

The production of agricultural crops in any environment or cropping system is ultimately the result of the biomass produced and the magnitude of that partitioned biomass for the harvested organ, which is measured in terms of biomass parity by the harvest index (HI).

In simple terms, production is the result of the interaction between the genetic potential of a cultivar and the biotic and abiotic factors that reduce that genetic potential. At the field level, plants are continuously subjected to multiple interactions with favorable results during most of the productive cycle, due to the plant's ability to adapt quickly to variations in soil and climate conditions and to technological support through agricultural practices.

The soybean plant is organized on the main stem on which the lateral branches and internodes are formed where flower clusters are formed (**Figure 1**). The order of the branches and bunches on the main stem are listed according to their ontogenetic chronology.

During the soybean cultivation cycle, five ontogenetic stages are distinguished, which are important in the determination of yield, which are: (1) The formation of organs responsible for the fixation of CO_2 and the absorption of water and nutrients (leaves and roots). (2) The formation of potential harvest organs (pods racemes in lateral branch or main stem racemes). (3) The determination of the effective density of harvest organs (number of pods/raceme-plant). (4) The filling of the harvest organs (number of filling seeds/pod, weight seed/pod). (5) Loss of functionality of leaves and roots (vegetative organs senescence, mature seeds in the pods) [41].

These stages develop successively with a degree of mutual overlap that varies with the type of cultivar and the environmental conditions of growth [41–44].

Two components of production are essential in the determination of soybean production. The number of grains per plant and the weight of the grains. The number of grains per plant is more closely associated with yield and is the most sensitive to the influence of the environment. This depends on the morphogenesis of reproductive structures on top of which are formed as are branches and clusters of the main trunk [41, 45–47].

Ontogenetically, the number of flowers in soy largely exceeds the potential capacity for fixation, even under restrictive environmental conditions. The fixation of the grains depends on the fixation of the pods and this characteristic is very sensitive to the availability of resources, so any physiological stress during the fixation of the pods determines the levels of pod abortion, consequently of the grain potential [41, 45–47]. However, between flowering and fully developed pod (R4) or start grain filling (R5) there may be compensation between yield components, fewer pods compensate with an increase in the number of grains per pod and/or grain weight.

Thus, in studies on the plant effects of high concentrations of CO_2 and O_3 it will be important to define the density of plants in the experimental field that avoid

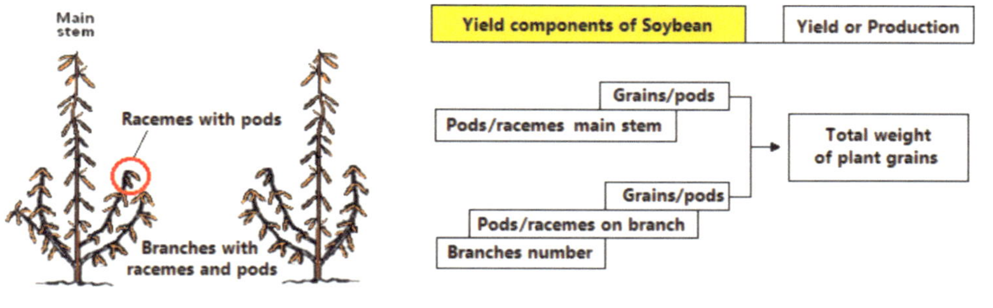

Figure 1.
Plant structure scheme with the branches and racemes contain pods, and the yield components in soybean which determine the production per plant.

high intraspecific competition during flowering and fixation of pods. An excessive competition for light can alter yield components such as number of branches and pods [34, 48, 49].

Adequate availability of light, temperature, water and nutrients during the period of performance determination, between the start of the pod formation (R3) and full green grain (R6) guarantees a high number of grains per plant [48, 50, 51, 52, 53]. It should be remembered that the plasticity in the number of grains per pod in soybeans is very low between 2.1–2.5 seed/pod [54, 55, 56], same with different cultural practices [57, 58, 59]. Thus, the number of grains per pod of soybeans can be less sensitive to stress compared to the number of pods.

The weight of grains, as the second component of the most sensitive yield, depends on the genotype and the environmental conditions that determine the photosynthesis capacity of the canopy, the translocation of assimilates, the duration of the filling stage and the competition between pods, and among grains on the same pod (source/sink ratio) [46].

The filling of the grain is strongly influenced by the availability and translocation of photoassimilates during the end of soybean development, before start of grain maturation (R7) [41]. Stresses by water deficit, thermal regime (below 25°C and above 35°C) can reduce the Leaf Area Index (LAI), thus like occurrence of rust and chronic exposures to O_3 can reduce the availability and use of photoassimilates by causing early senescence of the foliar area, and decreasing the photosynthesis and assimilated production [60, 61]. Thus, the benefits of the increase in CO_2 in the high photosynthesis and the more leaf area per plant can be decreased by foliar damage. Consequently, stresses during the start pod formation and the grain green full (stages R3-R6) affect the determination of the number of grains. Plant stresses between grain green full and start of grain maturation (stages R6–R7), decrease the weight of the grains on pods. On the other hand, a greater photosynthetic response in C3 plants such as soybeans to CO_2 increase, or high rates of photosynthesis among soybean cultivars, may not necessarily mean significant increases in production due to environmental interactions in the field, the possible effects of photosynthetic acclimatization, the increase of photorespiration by the increase in temperatures [62], and mainly by harvest index variability of soybean [63].

Climate changes, in particular, the increase in temperature and the concentrations of CO_2 and O_3 affect the development patterns and characteristics of the canopy as leaf area index (LAI) and internal structure of canopy [63, 64]. The magnitude of the alterations will be proportional to the environmental sensitivity of the cultivars and the applied productive management. Cultivars less sensitive to the indicated factors, the re-adaptation of population densities and arrangements in plant spacing may be the most immediate strategy such as resilience to climate change.

2.3 Changes in plant height, branches and racemes

Discussion on plant architecture is fundamental due to its link with the distribution of carbon allocation. Moreover, the understanding of plant shape allows for identification of plant features which are more strongly affected by environmental conditions such as CO_2 and weather parameters. Plants grown with higher CO_2 are taller than the plants in present conditions in the most several cultivars of soybean due to more nodes [34, 65, 66]. However, it is also possible to find no stimuli for the increase in the final height of the plant, which may be more likely in cultivars of certain growth habit [34]. Early results of [67] showed increase in height of the plant directly related with increase of more internodes and length of branches, or both.

The elevated temperature regime under CO_2 enrichment influences the growth and development of soybean plants. The temperature in plants, mainly affecting cell division, elongation rates, metabolic rates of photosynthesis and respiration in the daily cycle [68–71].

A mean temperature range of 29-31°C has been indicated as the optimum range for soybean vegetative growth [11, 12]. The main stem plastochron interval decreases and the final main stem node number increases in soybean with higher [CO_2] (660 μmolmol^{-1}) accompanied by a rise in mean temperature between 22.5 and 32.5°C [72]. Thus, the temperature increase can be favorable if closed to optimal temperature to soybean growth, even in conditions of greater availability of photoassimilates under higher concentrations of CO_2. Statistical analysis of correlation carried out by [34, 65] demonstrated a positive correlation between the plant height and the number of racemes, cultivar-dependent, under elevated CO_2 with 750 and 548 μmol.mol^{-1} and air temperatures, respectively. There was positive and significant increases in the number of nodes of soybean plant grown with elevated [CO_2]. It can be argued that the interaction between elevated [CO_2] and temperature in soybean influences the plant weight in two different ways, being through more number of internodes or internode length as also observed by [73] in 700 μmol.mol^{-1}. But depends on the on the cultivar response. It is important to note that changes in the plant height should induce modification in the configuration of other plant components, such as branches, pods and racemes.

The number of branches, racemes and total pods are the most important components for yield, and exhibit the highest correlation with the total yield [57, 74, 75]. A reduction by 18.5% in CO_2 enrichment in sensible cultivar to high CO_2 plant responses, and similar tendency happened in the number of pods/branch and grains/branch, with 35.1% and 35.2% decrease [34]. However, the number of grains/pods on branches can be remained unaltered in modern cultivars with small canopy or increase [66]. Based on previous studies one may anticipate that the increase of [CO_2] and warmer conditions may not contribute to the increasing yield due to reduction in the number of pods and grains on branches, if the current spacing and plant density remain unchanged. High yields were found with increased spacing under elevated CO_2 concentration [76]. In the evaluations of branch ontogeny or length of branches, the greatest branching plasticity of the US cultivars compared with Japanese cultivars should also be considered, together the inverse relation between the total length of branches and the density of plants [77].

The number of the racemes per plant and the number of pods or grains per racemes also respond to higher CO_2 depending on the cultivar. An increase in the number of the racemes/plant by 27% and 35% in grains/raceme in most sensitive of two cultivars, named Conquista [34]. In insensible modern cultivar (most ambient stability genotype) were not different in these characteristics. Additionally, there is a positive correlation between plant height and the number of racemes in sensible cultivar (Conquista) ($r = 0.67$; $P \leq 0.005$) [34].

A higher number of racemes lead to more pods and grains, which implies that a higher number of racemes in the main stem could partially compensate for the loss of pods and grains by the absence of lateral branches. The number of seeds per pod in soybean is very stable characteristic and can vary between 2.1 and 2.5 [54, 57, 58, 78] thus, the increase of the genetic plasticity of this characteristic can be a way for the increase of the productivity per unit of plant and area. Actually, this characteristic, perhaps, is the most limiting to increase the production in present and future environment.

The reduction in the number of branches observed in sensible cultivars under high competition among plants promoted by the higher CO_2 concentration can inhibit the axillary buds ontogeny in early vegetative stages [34, 55]. There is a need for further studies to elucidate the mechanisms of inhibition of branch ontogeny, and how early foliar self-shading can influence the ontogeny of branches and the number of its internodes.

According to the meta-analysis performed by [37], the increase of CO_2 in the growing environment results in a 35% increase in total dry matter/plant and the total leaf area/plant between 18 and 25% in soybean. This increase may result in larger dimensions of the canopy and the early occurrence of shading in the lower region of the canopy negatively affecting not only the ontogeny of the branches, but also the number of flowers and pods.

2.4 Changes in pods, grains and the grain weight

The pod sets, is the most variable yield component after the branch number. The integrated changes to plant level, such as total pods and grains per plant have been shown to cause the differences among plants under CO_2 enrichment.

The pod number per plant increases around 14% [37]. Previously, continuous and significant increases in the number of pods were found by [39] with increasing day/night thermal regime (18/12, 22/16 and 26/20°C) in interaction with each [CO_2] ranging from 350 (control), 650 and 1000 ppm of CO_2, respectively, in the cultivar Ransom cultivated under non-limiting conditions of water, nutrients and light inside a phytotron. Is evident the increase of the number of pod in all the racemes orders when the [CO_2] is near of 700 ppm [34, 67], however, the intensity of response is cultivar-dependent [34]. An evaluation about the relative partition of pods and grains per plant showed a greater relative partition of pods and grains in the first (basal) branches and in the first nodes of the branches and smaller relative partition in the subsequent branches and nodes [34].

In this way the gains in the first branches and nodes are lost logically by the reduction of pods and grains in the subsequent positions. Thus, a compensatory effect is established that cancels the initial gain, which may explain the small increase in production (7%) [34]. These authors, concluded that the ontogenic changes with respect to the formation of a smaller number of branches may be the cause of the low production gains under the effect of high [CO_2] due to the early self-shading.

Continuous shade between 60 and 90%, from initial bloom reduced pods per plant between 34 and 78%, respectively [79]. Several previous studies have found a reduction in the number of pods as the main factor of self-shading in soybean [80–82]. Additionally, [83] showed a greater sensitivity to shading of the number of pods per plant compared with the number of main stem nodes and the number of branches in two soybean cultivars grown with 50% of shading during soybean flowering. These authors [83] also verified increases in flower and pod abortion when the shading occurred together with lower temperatures, like 18°C day/10°C night.

The broader analysis of the grain weight in yield of the soybean points this characteristic to the low contribution to gain a significant increase in soybean production under conditions of high CO_2 concentration [37], despite having shown that the increase of the weight of the grains is possible in several cultivars [34, 65, 67].

Increases in the weight of grains have also been reported in soybean plants grown in an environment with elevated CO_2, independent of the changes in thermal regime even below the optimal growth temperature [38, 84]. There was strong fall in the seed weight in thermal regimens above the optimal temperature 32/22°C day/night, and increase by up to 13.5% in the grain weight in modern and landrace types when the temperatures during grown season closed to optimal temperature of soybean production [34, 46, 52, 85]. The increasing in weight of each grain is possible due to the existence of large genetic variability soybean species [86]. Besides the thermal regime influence, the long-term exposures of soybean plants to elevated [CO_2] can also change source-sink relations, and grain filling [76, 87].

The ratios of seed mass per plant, measured as seed mass obtained in elevated CO_2 compared the seed mass per plant in ambient (current CO_2 concentration), found by [34], was 1.13 for two contrast cultivars in canopy structure and size, modern (small canopy) and ancient (big canopy), and these ratios was coherent with the range from 0.93 to 1.87 previously reported by [84]. It has also been verified that higher [CO_2] and favorable temperature regime increase the grain weight through enhancements of sink-force of grains [34, 88]. However, the question remains whether this increase in sink-force is the same in all grains regardless of the position they occupy in the soybean plant. The variation in the number of pods and grains within their position in the branches and racemes helps explain how and where in the plant the changes occurred in relation to treatments. The number of grains per pod although it has lower variability has high heritability and greater positive effect on production [89].

2.5 Conclusions

The increase of [CO_2] in the soybean growing environment should lead to increases of among 7% [34] to 40% [37, 90, 91, 92, 93], and this maximum will depend on how the yield components are affected during growing season, which will depend on the cultivar, the density used and the interaction with temperatures close to the optimal temperature range for a particular cultivar. Temperatures that exceed the range of the optimal temperatures, cause negative alterations of the production of biomass and affect the partition for the formation of grains on the branches and clusters, reducing the yield. The elevated [CO_2] (around of 750 ppm) will attenuate the negative effects of the highest air temperatures because the carboxylase activity of Rubisco (photosynthetic enzyme of C3 plants) is favored by the higher internal concentration of CO_2 in the sub-stomatal chamber, resulting in photosynthetic rates higher than those obtained in the current CO_2 concentration.

The yield components most sensitive to the increase in atmospheric [CO_2] are the number of lateral branches, number of racemes in the main stem [34, 77, 79]. The number of pods formed will depend on the number on productive nodes formed on branches and in the main stem. Thus, plants adapted to future conditions should be able to maintain a high number of productive nodes per plant. The mechanisms of inhibition of the ontogeny of branches, mainly of the basal ones, still have to be explored, to define strategies of improvement, or management of densities between plants. The number and weight of the grains appear to be the most stable, which means that the increments of the production of soy may depend totally on the number of pods formed by the lateral branches.

Alteration in specific yield components and source-sink relationship is common in elevated CO_2, mainly in warmer climate at the level of branches, pods and grains. A better understanding of the response of soybean cultivars production, or for genotype screening, requires an evaluation of yield components, mainly in the branch level. Such is necessary to improve our understanding of sensible yield components in soybean genotypes and their ability to tolerate the impacts of the future climates.

Avoid intra-specific competition in future scenario by CO_2-increases, implies the need to avoid negative effects of intra-specific competition, such as self-shading [77, 79, 83], or the development of modern cultivars with narrow canopy and short branches. Thus, the current trend of the breeding soybean programs to decreasing the size of canopy [94, 95] as strategy to improve the productivity may continue to be the best strategy, even more so if plants were considered larger number of short branches to reduce the competition effect between the proximal and distant pods of the main stem as evidenced by [34].

Acknowledgements

The authors thank the National Postdoctoral Program CAPES-PROAP/BRAZIL, CiXPAG—Interaction of Climate Extremes, Air Pollution and Agro-ecosystems, coordinated by CICERO Climate Centre, UiO/MET Norway, Oslo, Norway, and by the Fundação de Amparo à Pesquisa do Estado de Minas Gerais—FAPEMIG (PCE-00891-17).

Conflict of interest

The authors declared that this chapter has no conflict of interest.

Author details

Milton E. Pereira-Flores and Flávio B. Justino*
Agricultural and Environmental Engineering Department DEA/UFV, Viçosa Federal University, Viçosa, MG, Brazil

*Address all correspondence to: fjustino@ufv.br

References

[1] Houghton JT, Ding Y, Griggs DJ, Noguer N, Linden PJVD, Dai X. Climate Change 2001: The Scientific Basis. Working Group I to the Third Assessment Report of the Intergovernmental Panel on Climate Change. Cambridge, UK: Cambridge University Press; 2001. 83 p

[2] IPCC Intergovernmental Panel on Climate Change. Climate Change. The Physical Science Basis. Working Group I. Fifth Assessment Report. Summary for Policymakers. WMO; 2013. 27 pp

[3] Constable GA, Rose A. Variability of soybean phenology response to temperature, daylength and rate of change in daylength. Field Crops Research. 1988;**18**(1):57-69. DOI: 10.1016/0378-4290(88)90059-7

[4] Câmara GMS, Sediyama T, Dourado-Neto D, Bernardes MS. Influence of photoperiod and air temperature on the growth, flowering and maturation of soybean (*Glycine max* (L.) Merrill). In: Presented at 4th Congress of the European Society for Agronomy, 7-11 July, 1996. Veldhoven-Wageningen. The Netherlands: Publish in Sci.agric; Piracicaba54(Numero Especial); 1997. pp. 149-154. http://www.scielo.br/pdf/sa/v54nspe/17.pdf

[5] Wu T, Li J, Wu C, Sun S, Mao T, Jiang B, et al. Analysis of the independent- and interactive-photo-thermal effects on soybean flowering. Journal of Integrative Agriculture. 2015;**14**(4):622-632. DOI: 10.1016/S2095-3119(14)60856-X

[6] Rockenbach AP, Otomar Caron B, Queiróz de Souza V, Elli EF, Machado de Oliveira D, Monteiro GC. Estimated length of soybean phenological stages. Semina: Ciências Agrárias. 2016;**37**:1871. DOI: 10.5433/1679-0359.2016v37n4p1871

[7] Avila AMH, Farias JRB, Pinto HS, Pilau FG. Climatic Restrictions for Maximizing Soybean Yields. In: Board JE, editor. A Comprehensive Survey of International Soybean Research - Genetics, Physiology, Agronomy and Nitrogen Relationships. Rijeka: Intech press; 2013. pp. 367-375

[8] Sinclair TR, Neumaier N, Farias JRB, Nepomuceno AL. Comparison of vegetative development in soybean cultivars for low-latitude environments. Field Crops Research. 2005;**92**:53-59

[9] Schoffel ER, Volpe CA. Contribuição relativa da temperatura do ar no desenvolvimento de três cultivares de soja. Revista Brasileira de Agrometeorologia. 2002;**10**(2):97-104

[10] Empresa Brasileira De Pesquisa Agropecuária, Centro Nacional de Pesquisa da Soja. Tecnologias de produção de soja: região central do Brasil 2012/2013. Londrina; 2011. 76 p

[11] Setiyono TD, Cassman KG, Specht JE, Dobermann A, Weiss A, Yang H, et al. Simulation of soybean growth and yield in near-optimal growth conditions. Field Crops Research. 2010;**119**:161-174

[12] Ziska LH, Bunce JA. Growth and photosynthetic response of three soybean cultivars to simultaneous increases in growth temperature and CO_2. Physiologia Plantarum. 1995;**94**:575-584

[13] Kumagai E, Tacarindua CP, Homma K, Shiraiwa T, Sameshima R. Effects of elevated CO_2 concentration and temperature on seed production and nitrogen concentration in soybean (*Glycine max* (L.) Merr.). Journal of Agricultural Meteorology. 2012;**68**(1):1-13

[14] Jones JW, Boote KJ, Jagtap SS, Mishoe JW. Soybean development. In: Hanks J, Ritchie JT, editors. Modeling Soil and Plant Systems. Madison, WI: American Society of Agronomy; 1991. pp. 71-90

[15] Avila AMH, Farias JRB, Pinto HS, Pilau FG. Climatic restrictions for maximizing soybean yields. In: Board JE, editor. A Comprehensive Survey of International Soybean Research - Genetics, Physiology, Agronomy and Nitrogen Relationships. Rijeka: Intech Press; 2013. pp. 367-375

[16] Hesketh JD, Myhre DL, Willey CR. Temperature control of time intervals between vegetative and reproductive events in soybeans. Crop Science. 1973;**15**:250-254

[17] Boote KJ, Jones JW, Hoogenboom G. Simulation of crop growth: CROPGRO model. In: Peart RM, Curry RB, editors. Agricultural Systems Modeling and Simulation. New York: M. Dekker; 1998. pp. 651-691

[18] Pandey JP, Torrie JH. Path coefficient analysis of seed yield components in soybeans (*Glycine max* (L.) Merr.). Crop Science. 1973;**13**(5):505-507

[19] Swank JC, Egli DB, Pfeiffer TW. Seed growth characteristics of soybean genotypes differing in duration of seed fill. Crop Science. 1987;**27**(1):85-89

[20] Thuzar M, Puteh AB, Abdullah NAP, Mohd. Lassim MB, Jusoff K. The effects of temperature stress on the quality and yield of soya bean [(*Glycine max* L.) Merrill.]. The Journal of Agricultural Science. 2010;**2**(1):172-179

[21] Onat B, Bakal H, Gulluoglu L, Arioglu H. The effects of high temperature at the growing period on yield and yield components of soybean [*Glycine max* (L.) Merr] varieties. Turkish Journal of Field Crops. 2017;**22**(2):178-186. DOI: 10.17557/tjfc.356210

[22] Kantolic AG, Slafer GA. Photoperiod sensitivity after flowering and seed number determination in indeterminate soybean cultivars. Field Crops Research. 2001;**72**:109-118

[23] Kantolic AG, Slafer GA. Development and seed number in indeterminate soybean as affected by timing and duration of exposure to long photoperiods after flowering. Annual Botany. 2007;**99**:925-933

[24] Major DJ, Johnson DR, Tanner JW, Anderson IC. Effects of daylength and temperature on soybean development. Crop Science. 1975;**15**(2):174-179. DOI: 10.2135/cropsci1975.0011183X001500020009x

[25] Huxley PA, Summerfield RJ, Hughes AP. The effect of photoperiod on development of soybean and cowpea cultivars grow in the U.K. in summer. Experimental Agriculture. 1974;**10**:225-239

[26] Destro D, Carpentieri-Pípolo V, Kiihl RAS, Almeida LA. Photoperiodism and genetic control of the long juvenile period in soybean: A review. Crop Breeding and Applied Biotechnology. 2001;**1**(1):72-92

[27] Watanabe S, Harada K, Abe J. Genetic and molecular bases of photoperiod responses of flowering in soybean. Breeding Science. 2012;**61**:531-543

[28] Acock MC, Acock B. Photoperiod sensitivity during soybean flower development. Biotronics. 1995;**24**:25-34

[29] Justino F, Oliveira EC, Rodrigues RA, Gonçalves PHL, Souza PJOP, Stordal F, et al. Mean and interannual variability of maize and soybean in brazil under global warming conditions.

American Journal of Climate Change. 2013;**2**:237-253

[30] Embrapa Soja. Empressa Brasileira de Pesquisa Agropecuaria. Tecnologias de produção de soja região central do Brasil 2009 e 2010: relatório do ano de 2008. Londrina: Embrapa Soja; 2008. 262 p. (Embrapa Soja. Sistemas de produção, 13)

[31] Constable GA, Rose A. Variability of soybean phenology response to temperature, daylength and rate of change in daylength. Field Crops Research. February 1988;**18**(1):57-69. DOI: 10.1016/0378-4290(88)90059-7

[32] Penalba OC, Bettolli ML, Vargas WM. The impact of climate variability on soybean yields in Argentina. Multivariate regression. Meteorological Applications. 2007;**14**:3-14. DOI: 10.1002/met.1

[33] Morgan PB. Soybean's future: Photosynthesis sucrose transport, dry mass accumulation and yield in a changing atmosphere. PhD thesis. Urbana-Champaign, IL, USA: University of Illinois; 2004

[34] Pereira-Flores ME, Justino F, Ruiz-Vera UM, Stordal F, Melo AAM, Rodrigues RA. Response of soybean yield components and allocation of dry matter to increased temperature and CO_2 concentration. AJCS. 2016;**10**(6):808-818. DOI: 10.21475/ajcs.2016.10.06.p7310

[35] Ruiz-Vera UM, Siebers M, Gray SB, Drag DW, Rosenthal DM, Kimball BA, et al. Global warming can negate the expected CO_2 stimulation in photosynthesis and productivity for soybean grown in the Midwestern United States. Plant Physiology. 2013;**162**(1):410-423

[36] Nakamoto H, Zheng SH, Tanaka K, Yamazaki A, Furuya T, Iwaya-Inoue M, et al. Effects of carbon dioxide enrichment during different growth periods on flowering, pod set and seed yield in soybean. Plant Production Science. 2004;**7**(1):11-15. DOI: 10.1626/pps.7.11

[37] Ainsworth EA, Davey PA, Bernacchi CJ, Dermody OC, Heaton EA, Moore DJ, et al. A meta-analysis of elevated [CO_2] effects on soybean (*Glycine max*) physiology, growth and yield. Global Change Biology. 2002;**8**:695-709

[38] Heinemann AB, Maia AD, Dourado-Neto D, Ingram KT, Hoogenboom C. Soybean (*Glycine max* (L.) Merr.) growth and development response to CO_2 enrichment under different temperature regimes. European Journal of Agronomy. 2006;**24**:52-61

[39] Sionit N, Strain BR, Flint EP. Interaction of temperature and CO_2 enrichment on soybean: growth and dry matter partitioning. Canadian Journal of Plant Science. 1987;**67**(1):59-67. DOI: 10.4141/cjps87-007

[40] Bishop KA, Betzelberger AM, Long SP, Ainsworth EA. Is there potential to adapt soybean (*Glycine max* Merr.) to future [CO_2]? An analysis of the yield response of 18 genotypes in free-air CO_2 enrichment. Plant, Cell & Environment. 2015;**38**(9):1765-1774. DOI: 10.1111/pce.12443. Epub 2014 Oct 27

[41] Hall AJ. Los componentes fisiológicos del rendimiento de los cultivos. Revista de la Facultad de Agronomía. 1980;**1**(1):73-86

[42] Fehr WR, Caviness CE. Stages of soybean development. Ames: Iowa State University, (Special Report, 80); 1977. 12 p

[43] Ramesh P, Gopalaswamy N. Heat unit requirement and prediction of developmental stages in soybean. Journal of Agronomy and Crop Science. 1991;**167**:236-240. DOI: 10.1111/j.1439-037X.1991.tb00869.x

[44] Aykuz FA, Kandel H, Morlock D. Developing a growing degree day model for North Dakota and Northern Minnesota soybean. Agricultural and Forest Meteorology. 2017;**239**:134-140

[45] Pedersen P. Soybean Growth and Development. Department of Agronomy. Iowa, USA: Iowa State University Extension and Outreach; 2007. 28 p

[46] Liu B, Liu XB, Wang C, Li YS, Jin J, Herbert SJ. Soybean yield and yield component distribution across the main axis in response to light enrichment and shading under different densities. Plant, Soil and Environment. 2010;**56**(8):384-392

[47] Kato S, Fujii K, Yumoto S, Ishimoto M, Shiraiwa T, Sayama T, et al. Seed yield and its components of indeterminate and determinate lines in recombinant inbred lines of soybean. Breeding Science. 2015;**65**(2):154-160. DOI: 10.1270/jsbbs.65.154

[48] Mathew JP, Herbert SJ, Zhang S, Rautenkranz AAF, Litchfield GV. Differential response of soybean yield components to the timing of light enrichment. Agronomy Journal. 2000;**92**:1156-1161

[49] Kokubun M. Physiological Mechanisms Regulating Flower Abortion in Soybean. In: Ng T-B, editor. ISBN: 978-953-307-219-7. Soybean - Biochemistry, Chemistry and Physiology. Rijeka: InTech; 2011, Available from: http://www.intechopen.com/books/soybean-biochemistry-chemistry-and-physiology/physiologicalmechanisms-regulating-flower-abortion-in-soybean

[50] Kumar A, Pandey V, Shekh AM, Kumar M. Growth and yield response of soybean (*Glycine max* L.) in relation to temperature, photoperiod and sunshine duration at Anand, Gujarat, India. American-Eurasian Journal of Agronomy. 2008;**1**(2):45-50

[51] Liu X, Herbert SJ, Hashemi AM, Litchfield GV, Zhang Q, Barzegar AR. Yield and yield components responses of old and new soybean cultivars to source-sink manipulation under light enrichment. Plant, Soil and Environment. 2006;**52**(4):150-158

[52] Appiah AK, Helget R, Xu Y, Jixiang W. Response of soybean yield and yield components to phosphorus fertilization in South Dakota. Conference on Applied Statistics in Agriculture; 2014. DOI: 10.4148/2475-7772.1001

[53] Rodrigues TR, Casaroli D, Evangelista AWP, Júnior JA. Water availability to soybean crop as a function of the least limiting water range and evapotranspiration. Pesquisa Agropecuária Tropical Goiânia. 2017;**47**(2):161-167. DOI: 10.1590/1983-40632016v4743746

[54] Peluzio JM, Vaz-de-Melo A, Afférri FS, Silva RR, Barros HB, Nascimento IR, et al. Variabilidade genética entre cultivares de soja, sob diferentes condições edafoclimáticas. Pesquisa Aplicada & Agrotecnologia. 2009;**2**(3):21-29

[55] Soares MM, Oliveira GL, Soriano PE, Sekita MC, Sediyama T. Performance of soybean plants as function of seed size: II. Nutritional stress. Journal of Seed Science. 2013;**35**(4):419-427. DOI: 10.1590/S2317-15372013000400002

[56] Nico M, Mantese AI, Miralles DJ, Kantolic AG. Soybean fruit development and set at the node level under combined photoperiod and radiation conditions. Journal of Experimental Botany. 2016;**67**(1):365-377. DOI: 10.1093/jxb/erv475

[57] Pires JLF. Variabilidade espacial dos componentes de produção de plantas de soja em comunidade. Tese (Doutorado em Fitotecnia). Porto Alegre: Faculdade

de Agronomia, Universidade Federal do Rio Grande do Sul; 2002. 139 p

[58] Dalchiavon FC, Passos e Carvalho M. Correlação linear e espacial dos componentes de produção e produtividade da soja linear and spatial correlation of the yield components and soybean yield. Londrina, Semina: Ciências Agrárias, Londrina. 2012;**33**(2):541-552

[59] Fageria NK, Moreira A, Cardoso M, Moraes M. Influence of lime and gypsum on yield and yield components of soybean and changes in soil chemical properties. Communications in Soil Science and Plant Analysis. 2014;**45**. DOI: 10.1080/00103624.2013.861906

[60] Moreira EN, Vale FXR, Paul PA, Rodrigues FA, Jesus WC. Temporal index in soybean cultivars of different maturity groups. Plant Disease. 2015;**99**:1216-1226

[61] Dermody O, Long SP, DeLucia EH. How does elevated CO_2 or ozone affect the leaf-area index of soybean when applied independently? New Phytologist. 2006;**169**:145-155

[62] Bloom AJ. As carbon dioxide rises, food quality will decline without careful nitrogen management. California Agriculture. 2009;**63**(2):67-72

[63] Krisnawatia A, Adie MM. Variability of biomass and harvest index from several soybean genotypes as renewable energy source. Energy Procedia. 2015;**65**:14-21

[64] Gray SB, Dermody O, DeLucia EH. Spectral reflectance from a soybean canopy exposed to elevated CO_2 and O_3. Journal of Experimental Botany. 2010;**61**(15):4413-4422. DOI: 10.1093/jxb/erq244

[65] Castro JC, Dohleman FG, Bernacchi CJ, Long SP. Elevated CO_2 significantly delays reproductive development of soybean under Free-Air Concentration Enrichment (FACE). Journal of Experimental Botany. 2009;**60**(10):2945-2951

[66] Jin J, Li Y, Liu X, Wang G, Tang C, Yu Z, et al. Elevated CO_2 alters distribution of nodal leaf area and enhances nitrogen uptake contributing to yield increase of soybean cultivars grown in Mollisols. PLoS One. 2017;**12**(5):e0176688. DOI: 10.1371/journal.pone.0176688

[67] Rogers HH, Cure JD, Thomas JF, Smith JM. Influence of elevated CO_2 on growth of soybean plants. Crop Science. 1984;**24**:361-366

[68] Bruhn D. Plant respiration and climate change effects. Ph.D. thesis Plant Research Department Botanical Institute University of Copenhagen. Risø National Laboratory, Roskilde. Dannmark April; 2002. 139 p

[69] Francis D, Barlow PW. Temperature and the cell cycle. Symposia of the Society for Experimental Biology. 1988;**42**:181-201

[70] Tardieu F, Granier C. Quatitative analysis of cell division in leaves: methods, developmental patterns and effects of environmental conditions. Plant Molecular Biology. 2000;**43**:555-567

[71] Tardieu F, Reymond M, Hamard P, Granier C, Muller B. Spatial distributions of expansion rate, cell division rate and cell size in maize leaves: a synthesis of the effects of soil water status, evaporative demand and temperature. Journal of Experimental Botany. 2000;**51**(350):1505-1514

[72] Baker JT, Allen LH, Boote KJ, Jones P, Jones JW. Response of soybean to air temperature and carbon dioxide concentration. Crop Science. 1989;**29**(1):98-105

[73] Allen LH, Zhang L, Boote KJ, Hauser B. Elevated temperature intensity, timing, and duration of exposure affect soybean internode elongation, mainstem node number, and pod number per plant. The Crop Journal. 2018;**6**:148-161. DOI: 10.1016/j.cj.2017.10.005 2214-5141

[74] Perini LJ, Júnior NSF, Destro D, Prete CEC. Componentes da produção em cultivares de soja com crescimento determinado e indeterminado. Semina: Ciências Agrárias, Londrina. 2012;**33**(1):2531-2544

[75] Egli DB. The relationship between the number of nodes and pods in soybean communities. Crop Science. 2013;**53**:1668-1676. DOI: 10.2135/cropsci2012.11.0663

[76] Kumagai E, Tacarindua CP, Homma K, Shiraiwa T, Sameshima R. Effects of elevated CO_2 concentration and temperature on seed production and nitrogen concentration in soybean (*Glycine max* (L.) Merr.). Journal of Agricultural Meteorology. 2012;**68**(1):1-13

[77] Yoshihira AT, Tatsuhiko Shiraiwa T. Branch development responses to planting density and yield stability in soybean cultivars. Plant Production Science. 2016;**19**(3):331-339. DOI: 10.1080/1343943X.2016.1157443

[78] OSU Ohio State University. 2012. Worksheet to Estimate Soybean Yield. Aug 21. http://cornandsoybeandigest.com/soybean/4-components-estimating-soybean-yield

[79] Egli DB. Flowering, pod set and reproductive success in soya bean. Journal of Agronomy and Crop Science. 2005;**191**:283-291

[80] Tanaka A, Fujita K, Tanaka Y. Effects of shading on nitrogen fixation and combined nitrogen absorption in soybean. Japanese Journal of Soil Science and Plant Nutrition. 1980;**51**(4):281-284

[81] Ishikawa T, Nakaseko K, Goto K. Effects of shading on various growth stage to determinate and indeterminate soybeans. Report of the Hokkaido Branch, the Japanese Society of Breeding and Hokkaido Branch, the Crop Science Society of Japan. 1984;**24**:21

[82] Jiang H, Egli DB. Shade induced changes in flower and pod number and flower and fruit abscission in soybean. Agronomy Journal. 1993;**85**:221-225

[83] Kurosaki H, Yumoto S. Effects of low temperature and shading during flowering on the yield components in soybeans. Plant Production Science;**6**(1):17-23. DOI: 10.1626/pps.6.17

[84] Hikosaka K, Kinugasa T, Oikawa S, Onoda Y, Hirose T. Effects of elevated CO_2 concentration on seed production in C_3 annual especies. Journal of Experimental Botany. 2011 Feb;**62**(4):1523-1530. DOI: 10.1093/jxb/erq401

[85] Thomas JMG, Boote K, Allen JLH Jr, Gallo-Meagher M, Davis JM. Elevated temperature and carbon dioxide effects on soybean seed composition and transcript abundance. Crop Science. 2003;**43**:1548-1557

[86] Mulato BM, Möller M, Zucchi MI, Quecini V, Pinheiro JB. Genetic diversity in soybean germplasm. Pesquisa Agropecuária Brasileira. 2010, 2010;**45**(3):276-283

[87] Isopp H, Frenher M, Long SP, Nösberger J. Sucrose-phosphate synthase responds differently to source-sink relations and to photosynthetic rates: *Lolium perenne* L. growing at elevated pCO_2 in the field. Plant, Cell & Environment (Oxford). 2000;**23**(6):597-607

[88] Lemoine R, Camera SL, Atanassova R, Dédaldéchamp F, Allario T, Pourtau N, et al. Source-to-sink transport of sugar and regulation by environmental factors. Frontiers in Plant Science. 2013;**4**:272. DOI: 10.3389/fpls.2013.00272

[89] Sudarić A, Vratarić M, Duvnjak T. Quantitative genetic analysis of yield components and grain yield for soybean cultivars. Agricultural Institute – Croatia. In: Kauffman HE editor. Proceedings of the World Soybean Conference VI, Chicago, USA. 479; 2002. 28 p. Vratarić, M., Sudarić, A. (2000.): [1] Soja. Book. Poljoprivredniinstitut Osijek, Osijek. pp. 25-58

[90] Ziska LH, Bunce JA, Caulfield FA. Rising atmospheric carbon dioxide and seed yield of soybean genotypes. Crop Science. 2001;**41**:385-391

[91] Li D, Liuc H, Qiaoa Y, Wang Y, Cai Z, Donga B, et al. Effects of elevated CO_2 on the growth, seed yield, and water use efficiency of soybean (*Glycine max* (L.) Merr.) under drought stress. Agricultural Water Management. 2013;**129**:105-112

[92] Li D, Liuc H, Qiaoa Y, Wang Y, Cai Z, Donga B, et al. Effects of elevated CO_2 on the growth, seed yield, and water use efficiency of soybean (*Glycine max* (L.) Merr.) under drought stress. Agricultural Water Management. 2013;**129**:105-112

[93] Hikosaka K, Kinugasa T, Oikawa S, Onoda Y, Hirose T. Effects of Elevated CO_2 Concentration on Seed Production in C_3 Annual Species. Journal of Experimental Botany. Feb 2011;**62**(4):1523-30. DOI: 10.1093/jxb/erq401

[94] Morrison MJ, Voldeng HD, Cober ER. Agronomic changes from 58 years of genetic improvement of shortseason soybean cultivars in Canada ECORC contribution no. 001482. Agronomy Journal. 2000;**92**:780-784. DOI: 10.2134/agronj2000.924780x

[95] Jin J, Liu X, Wang G, Mi L, Shen Z, Chen X, et al. Agronomic and physiological contributions to the yield improvement of soybean cultivars released from 1950 to 2006 in Northeast China. Field Crops Research. 2010;**115**:116-123

Chapter 3

Soybean Seed Production and Canopy Photosynthesis in Cultivation

Kiyoshi Nagasuga

Abstract

The mechanism of soybean seed production is very complicated. Soybean yield is strongly associated with pod number and seed number; these are prompted by light interception and growth during the period between beginning blooming and beginning seed. But vigorous shoot growth during the vegetative stage does not contribute to pod growth and harvesting. In humid regions of Asia, soybean cultivation is incorporated into the rotation cropping in converted paddy fields, and wet soil often causes poor germination. Soybean leaves, trifoliate wide flat leaves, are easy to concentrate to the upper layer of the canopy. This suppresses light penetration to the lower layer and, as a result, produces imperfect seed yield in spite of enough biomass. Daytime leaf movement is useful for light penetration and photoinhibition in leaf photosynthesis. Leaf photosynthesis is generally associated with high yield; however, the relationship between them is not clear. It is necessary for high soybean yield not only to elucidate the mechanisms that these factors suppress soybean seed production more clearly but also to select the cultivars and cultivation suitable for the climate in each area.

Keywords: biomass, canopy photosynthesis, cultivation, germination, light interception, lodging, pod, seed production

1. Introduction

Soybean is one of the important crops for oil and protein resources. Soybean production in 2016 is 336 million tons, and about 80% of the world production depends on a few major producers, the United States, Brazil and Argentina [1]. World soybean production continuously increases at a remarkable rate for the last several decades; it reached ca. 265 million tons in 2010 from ca. 30 million tons in 1970 [2]. This increase is associated with the improvement of soybean production; yield gain of the United States is 22.6 kg/ha/year from 1924 to 1997 and 12.1 kg/ha/year from 1950 to 1991 in China [3]. High soybean yield is also an important agricultural strategy not only in the major producers but also in the minor producers such as India, Japan and other Asian countries. Many researches about soybean production were actively made to achieve high yielding; for example, more than 60 years of researches have produced various types of soybean cultivars in Japan because soybean is the fundamental material of Japanese foods. However, these trials cannot find the breakthrough that improves soybean production in this country; Japanese

IntechOpen

soybean yield (160 kg/10a) has been steady values for the current decade [4]. World soybean production polarized.

Improvements in weed control and planting and harvest machinery as well as the incorporation of disease and lodging resistance into elite soybean germplasm have contributed to the yield improvement in continental North America [5]. In addition, many theories on the physiological traits associated with genetic yield improvement have been put forward; however, there are still some incongruencies among these theories [6–9]. Therefore, it is important for soybean yield improvement to understand the relationship between yield and quantitative physiological traits.

2. Seed yield and biomass

In general, the yield of crop plants can be expressed as the function of biomass and harvest index as follows:

$$\text{Yield} = \text{Biomass} \times \text{Harvest index}$$

Biomass is the total amount of living organisms in a given area. Biomass contains leaf, stem, root and seed (full maturity). Harvest index is harvested product (seed) as a percentage of total crop weight. Although high biomass does not always result in high yield, low biomass cannot result in high yield even if crop plants could realize remarkable high harvest index. So the increase in biomass is easier to result in high yield than that of harvest index. Meanwhile, the improvements of crop yield by breeding are also associated with the increase in harvest index, for example, 1.50 times increase in Chinese rice yield is attributed to 1.16 times that of biomass and 1.33 times that of harvest index by breeding [10]. It is depending on the crop species whether a numerous increase of crop yield is attributed to the increases in biomass, harvest index or both.

In soybean, harvest index is often regarded as seed/stem ratio, not seed/shoot ratio because soybean leaves start to get yellow around beginning maturity and have been shedded at full maturity. The reports on the role of harvest index on soybean seed yield include both significant [8] and no significant [6, 11–13] correlations with harvest index. However, even the researchers who reported significant correlation between yield and harvest index suggested that the increase in harvest index contributed little to increased yield potential of modern genotypes [9, 14]. It seems that harvest index is not an important contributor to genetic yield improvement; the increase in biomass is more useful for high yield of soybean plants.

Breeding often improves crop yield through the increase in biomass. Plant biomass is composed of 70–90% water and 10–30% dry matter [15]. Water is a very important material for life; this controls life maintenance and plant growth strongly. However, it is necessary for a large increase in plant biomass to get not only abundant water but also much dry matter. Dry matter is derived from CO_2 fixed through photosynthesis [16, 17]. Photosynthesis produces the basic carbohydrates used for producing more complex carbohydrates, proteins and lipids, all of which contribute to dry matter [18]. Dry matter weight is a more reliable measure of mass than fresh weight (biomass) because dry matter excludes the fluctuating water concentrations [19] so the effect of soybean biomass on yield is evaluated through that of total dry matter weight in this study.

Before the correlation between yield and total dry matter weight is mentioned, descriptions of growth stage of soybean during the growing season according to Fehr and Caviness (**Table 1**; [20]) and its relationship with growth parameters

Stage	Descriptions	
Vegetative stage		
VE	Emergence	Cotyledons above the soil surface
CV	Cotyledon	Cotyledons and unifoliates are fully expanded
V1	First node	One unrolled trifoliate (leaflets do not touch) on the main stem
V2	Second node	Two unrolled trifoliate (leaflets do not touch) on the main stem
V(n)	Nth node	N unrolled trifoliates (leaflets do not touch) on the main stem
Reproductive stage		
R1	Beginning bloom	One flower at any node on the main stem
R2	Full bloom	One flower at one of the two uppermost nodes on the main stem with a fully expanded trifoliate
R3	Beginning pod	A 0.5-cm-long pod at one of the four uppermost nodes on the main stem with a fully expanded trifoliate
R4	Full pod	A 2-cm-long pod at one of the four uppermost nodes on the main stem with a fully expanded trifoliate
R5	Beginning seed	Seed is 0.3 cm long in a pod at one of the four uppermost nodes on the main stem with a fully expanded trifoliate
R6	Full seed	A pod containing a green seed that fills the pod cavity at one of the four uppermost nodes on the main stem with a fully expanded trifoliate
R7	Beginning maturity	One pod anywhere on the main stem with the mature brown color
R8	Full maturity	95% of pods reached mature color

Table 1.
Descriptions of growth stage of Soybean [20].

including total dry matter weight are mentioned. Although soybean leaf is trifoliate, the cotyledon pulled by the hypocotyl emerges out of the soil (germination, VE), and the first leaves are two unifoliates (cotyledon, VC).

Infection of *Bradyrhizobium japonicum* and nodule formation start around the second node (V2). Plant length and leaf area index increase gradually after emergence and reach more than 80% maximum values around beginning bloom (R1). Canopy structure is also almost achieved around this period. On the other hand, total dry matter weight increases slowly from emergence to R1 and then exponentially until beginning pod (R3). The maximum values are reached at beginning seed (R5) or just before full seed (R6), and then total dry matter weight decreases by maturing. However, some cultivars with vigorous vegetative growth capacity sometimes increase total dry matter weight continuously around beginning maturity (**Figure 1A**) [21]. Soybean development is separated into vegetative period (emergence to R1) and reproductive period (R1–R7). However, vegetative growth (leaves, stems and nodes) extends from emergence to R5. The reproductive period is separated into flowering/pod formation period (R1–R6) and seed filling period (R5–R7). Pod and seed numbers are determined until R6 [22].

There is a hypothesis that soybean yield is limited by the activity of the source (i.e. canopy photosynthesis) [23]. One of the reasons is low ratio (19–64%) of pod set (from flower to pod) in soybean plants [24–26]; many flowers fall not only by environmental stress (i.e. low temperature [27], water shortage [28] and light energy shortage [29]) but also by nutrient competition between flowers in the plant

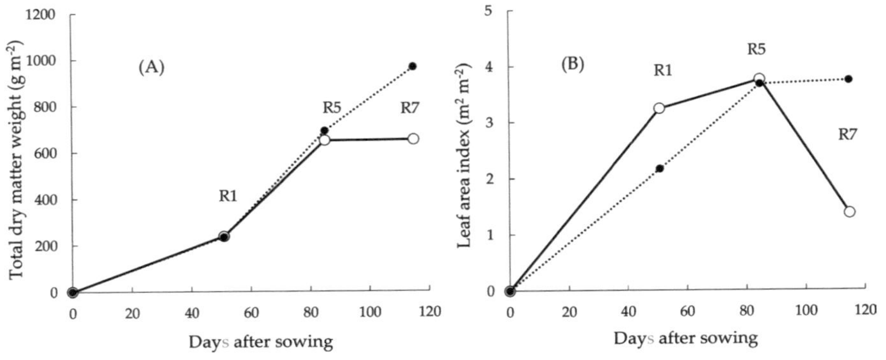

Figure 1.
Changes in total dry matter weight (A) and leaf area index (B) in Fukuyutaka (open) and Misato-zairai (close). Fukuyutaka and Misato-zairai are the maturity groupII determinate genotypes. Fukuyutaka, cultivated variety, has relatively staple seed yield; however Misato-zairai, local cultivar, has an unstable one [21].

or with vegetative organs [30]. Above-ground (particularly leaves) of crop plants is main photosynthetic organs and this increase is associated strongly with nitrogen content in the soil [31]. Basal fertilizer is generally applied to promote vigorous photosynthesis and crop growth from early growth stage. However, total dry matter weight before flowering has little effect on seed yield in soybean plants [32]. In addition, the reports on the association between biomass and seed yield have presented conflicting results; some researchers reported that total dry matter weight was not significantly related to yield improvement in 14 short-season soybean cultivars from different eras of release [8, 33, 34]. On the other hand, after the onset of reproductive development, many researchers indicate a positive association between dry matter accumulation and seed yield [9, 14, 35–37]. There are also conflict reports about more critical period after flowering, some reports described the period between R1 and R5 [36, 38, 39], and others are after R5 [35, 40]. However, seed weight is often associated with seed number and pod number strongly, and seed number depends on pod number. **Figure 2** shows variation in seed weight as a function of pod number (A), seed number (B) and the relationship between seed

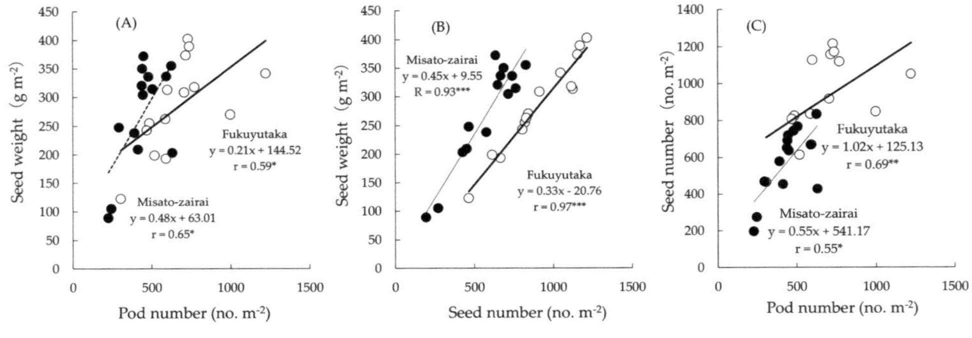

Figure 2.
*Variation in seed weight as a function of pod number (A), seed number (B) and the relationship between seed number and pod number (C) at maturity (R8) in Fukuyutaka (open) and Misato-zairai (close) cultivated at Mie Prefecture in Japan in 7 years (unpublished data). Experimental field, plant density, fertilization and the measurement of yield components in these experiments were the same as the previous studies [21, 41–44]. Main data are those of normal sowing (early-middle July sowing at Mie) from 2008 to 2014 (excluding 2012) [21, 41–44], and the same include early sowing (middle May and middle June in 2009) and irrigation treatment (from blooming in 2009 and from 1 month after sowing in 2013 and 2014) [43, 44]. r is correlation coefficient. *, **, ***: significant at 0.5, 0.1 and 0.01%, respectively.*

DOI: http://dx.doi.org/10.5772/intechopen.81808

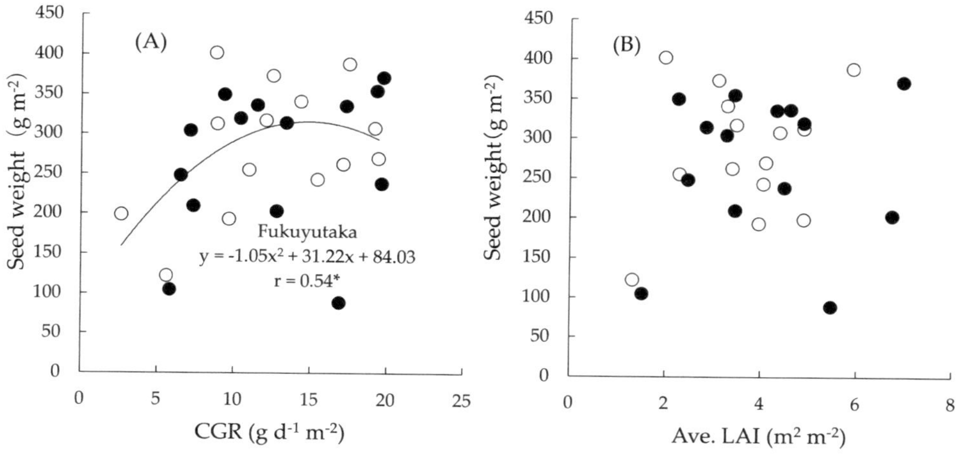

Figure 3.
Variation in seed weight at maturity (R8) as a function of crop growth rate (CGR, A) and average leaf area index (Ave. LAI, B) from beginning bloom (R1) to beginning seed (R5) in Fukuyutaka (open) and Misato-zairai (close) cultivated at Mie Prefecture in Japan in 7 years (2008–2014, unpublished data). The samples are the same as those in **Figure 2**, *and CGR and ave. LAI were measured according to the previous studies [21, 41, 43]. r is correlation coefficient.* *: *significant at 0.5%.*

number and pod number (C) at maturity (R8) in Fukuyutaka and Misato-zairai cultivated at Mie Prefecture in Japan in 7 years (2008–2014). Misato-zairai has larger seed size and lower seed number than Fukuyutaka; however, the relationships among seed weight, seed number and pod number are similar in these cultivars. Pod formation is nearly completed by R3 [39]. Although there are no positive relationships between seed weight and total dry matter weight at R5, crop growth rate (CGR, dry matter accumulation per day) during the period between R1 and R5 is easy to be associated with seed weight (**Figure 3**). The period between R1 and R5 is considered critical for the impact of assimilatory capacity on yield, and it would be necessary for high soybean yield to make active photosynthetic performance.

On the other hand, the initial growth often has a serious impact on seed yield in the monsoonal area of Asia. Soybean often cultivated in the converted paddy fields and excessive rainfall and poor drainage frequently depending on rotation cropping often cause waterlogging in this area. Soybean is vulnerable to waterlogging, which threatens soybean productivity [2]. The most sensitive growth stage is around emergence, particularly germination. Poor emergence has a serious impact on soybean production because this decreases plant density radically. Waterlogging suppresses soybean seed germination through the several manners. Firstly, waterlogging destroys seed tissue by abrupt cell expansion because soybean seeds absorb water rapidly in this condition [45, 46]. Secondly, waterlogging inhibits the respiratory activity of germinating seeds. Germination is an integrated process consisting of many metabolic events; therefore, active respiration is necessary for its metabolism and germination. Excessive soil moisture condition decreases oxygen concentration in the soil and inhibits seed respiration. Thirdly, soybean seeds in the waterlogging condition are vulnerable to soil-borne diseases, which often results in seed corruption and the inhibition of seed germination. On the other hand, waterlogging also suppresses soybean emergence physically. Waterlogging or hard rain destroys the crumb structure of culture soils, which makes soil surface hard like a plate. These suppress emergence through the blockage of hypocotyl extension even if seed germination succeeds. After emergence, waterlogging also threatens soybean productivity through the suppression of root respiration, growth and symbiotic nitrogen fixation by root nodules and the spread of soil disease infection. These have a great impact on vegetative growth, which inhibits pod number per plant, seed number per pod and seed weight. Meanwhile, the emergence is strongly correlated to plant density and hill number per area; suppression of emergence by waterlogging often decreases plant density dramatically. Because of insufficient shoot dry matter weight and leaf area per unit area, low plant density is difficult to result in high yield in various crop plants [47–49]. Therefore, low plant density by waterlogging has a serious impact on soybean yield. The cultural or genetic improvements of soybean emergence are more important in the monsoonal area of Asia.

3. Canopy photosynthesis and its related parameters

After emergence, active photosynthetic performance of canopy is useful for high seed production of soybean plants. Canopy photosynthesis can be composed of three parameters: leaf area index, light intercepting efficiency and photosynthetic rate per unit leaf area (leaf photosynthesis).

3.1 Leaf area index

Leaf area index means total leaf area per unit ground area [50]. Leaf area index can be expressed as the product of plant density and total leaf area of a plant. Plant density is a factor that is easy to control by cultivation: dense planting increased leaf area index through the increases in plant density easily and quickly. However, crop shoots in high plant density are spindly and often result in lodging [47–49]. So it is necessary for the increase in leaf area index both to keep optimum plant density and to increase leaves with the capacity of leaf appearance and expansion of crop plants itself.

Soybean leaves develop acropetally from individual nodal primordia on the main stem as well as on the branches. The rate of individual leaf appearance, expansion and duration as well as its interaction with environmental factors has an impact on leaf area index development. Each leaf unrolls, expands and persists in the expanded state for a period of time prior to senescence and finally abscission. **Figure 1B** shows change in leaf area index in Fukuyutaka and Misato-zairai. Typical changes in soybean leaf area index are that of Fukuyutaka: leaf area index increases dramatically during vegetative period, and maximum value is achieved by around R5. After that, leaf area index begins to decline and this decline accelerates by leaf abscission before R7. Leaf area index has an impact on canopy ability to intercept solar radiation and light interception up to 95% of incident solar radiation around R5 [5]. In this stage, canopy closure is observed and this corresponds to leaf area index of 3–5 [33]. A leaf area index greater than 5 can be sometimes observed; however, this is not easy to result in the increase in radiation interception any further [51].

High crop growth rate is usually associated with high leaf area index in crop plants [31]. Soybean leaf area controls strongly crop growth rate during the vegetative period, and the increase in total leaf area is associated with node number of main stem and specific leaf area (leaf area/leaf dry matter weight) [52]. Leaf growth is often associated with environmental factors (i.e. water, temperature, nitrogen, light intensity). During the vegetative period, leaf area increases with sufficient soil water, and severe soil drought restricts strictly leaf growth in the high air temperature condition; however, low air temperature condition negates the effect of soil water condition on leaf growth, and the difference in total leaf area between wet drought soil conditions is too small in Japanese soybean cultivar (**Figure 4**) [53].

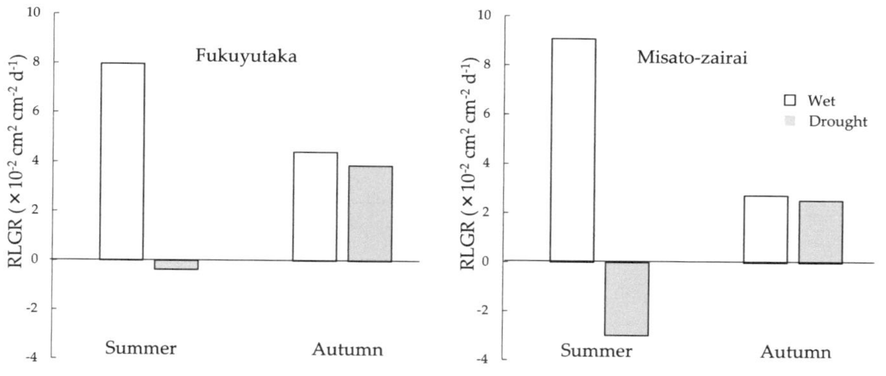

Figure 4.
Relative leaf growth rate (RLGR) of Fukuyutaka and Misato-zairai grown in wet and drought soil conditions during the vegetative stage in summer and autumn. After 1 month of cultivation, supply of water to some plants was suppressed for 2 weeks (drought). Relative growth rate is daily leaf area accumulation per leaf area [53].

There is an optimum leaf area index where crop growth rate reaches the maximum level. Crop growth rate increases with increment of leaf area index, attains the maximum level and then decreases with increase of leaf area index. In soybean plants, the increase in leaf area index is easy to result in that of crop growth rate during R1–R5 [41], and this is associated with higher specific leaf area, not leaf weight ratio (leaf dry matter weight/total dry matter weight) [41]. However, higher leaf area index has a negative impact on seed yield. There is a weak relationship between average leaf area index during the period between R1 and R5 and seed weight at R8 in Fukuyutaka and Misato-zairai (**Figure 3**). Higher leaf area index would be useful for high biomass, but not for seed production in soybean plants.

3.2 Light intercepting efficiency

Light extinction coefficient can be used to quantify the extent of light penetration into the plant canopy of various species. The light extinction within the canopy follows Lambert-Beer's law [31]:

$$\mathrm{Ln}\, I/I_0 = -kL$$

where I_0 is the incident light intensity on the canopy and I is the light intensity at a certain level in the canopy where the cumulative leaf area index from the top of the canopy is L. The value of k is the light extinction coefficient which is specific to the respective plant canopy. The decrease in the extinction coefficient represents an increase in light penetration into the plant canopy. A monocotyledonous plant (i.e. rice) showed a small value of k. The meristem of rice plant locates at the junction between the stem (leaf sheath) and root during the vegetative stage [54]. Although the meristem rises with making anthesis, much nodal primordia with long, thin and upright leaves still concentrate to the lower layer [54]. So, leaves are distributed widely from the lower to upper layer of the canopy (**Figure 5A**). On the other hand, a dicotyledonous plant with round leaves (i.e. soybean) showed larger k. The meristem of dicotyledonous plant locates at the shoot apex, and this rises continuously (or until flowering in determinate genotypes) [55]. Therefore, the nodal primordia locate at various layers of the canopy. In addition, the leaves expand from the apex of petiole. These result in the concentration of each leaf over the middle layer of the canopy (**Figure 5B**, [21]).

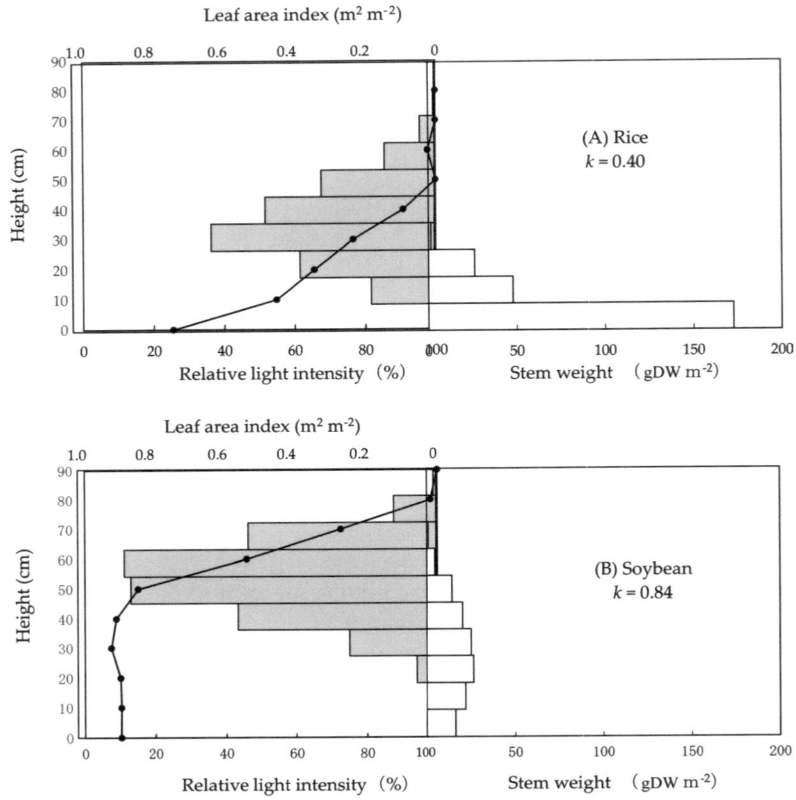

Figure 5.
Vertical distributions of leaf area index and stem dry matter weight (stem weight) of rice (A) and soybean (B) cultivated at Mie Prefecture in Japan just before flowering (original data). Soybean (Fukuyutaka) cultivation and the measurement of dry matter weight, leaf area and relative light intensity were mentioned in the previous study [21]. Young rice seedlings (Koshihikari) were transplanted to the paddy field at a hill spacing of 18 cm and a row spacing of 30 cm (18.5 hills m^{-2}) on 18 April 2016. Chemical fertilizers (N, P_2O_5 and K_2O) were applied at the rate of 4 g m^{-2} as basal and 1 g m^{-2} as top dressing on June 28. Measurements of the above parameters were made according to that of soybean [21]. The close circle represents relative light intensity inside the canopy. The stem includes leaf sheath (rice) or petiole (soybean). k represents light extinction coefficient.

The light intercepting efficiency (i.e. *k*) is strongly associated with morphological leaf traits and leaf distribution in the canopy [31]. Among them, leaf erectness is the most distinct determinant of light intercepting efficiency [31]; the improvement of leaf inclination angle by breeding often causes high seed yield in modern rice cultivars. In soybean, however, there is a report that light intercepting efficiency had no influence on canopy photosynthesis [56]. In comparison with Fukuyutaka, Misato-zairai has unstable seed yield, and un-erect and more dense leaves in the uppermost layer of the canopy were found in this cultivar [21]. However, there is no significant difference in light extinction coefficient between these two cultivars [21]. Although light intercepting efficiency may be useful for soybean seed production, the relationship among light intercepting efficiency, its related parameters and seed yield was complicated and unclear [21, 42].

3.3 Leaf orientation adjustment

High light intensity, such as full sunlight, is harmful to plant leaves because high light intensity induces photoinhibition in crop leaves [57]. In leguminous crop plants, including soybean, leaf orientation in the top layer of the canopy changes during the daytime (paraheliotropism, light avoidant movement [58]), which avoids receiving too much light and, as a result, photoinhibition. This mechanism appears to be turgor-mediated; paraheliotropism is controlled by turgor changes of the pulvinus tissue at the base of leaves [59]. Therefore, paraheliotropism is associated strongly with plant water stress; the degree of the midday avoidance becomes more pronounced in water-stressed plants as compared with well-watered ones [58, 60]. Paraheliotropic response to soil water availability is different between soybean cultivars. Midday leaf angle of the species which grow in the wet habitat is more vertical than that of the species which grow in the dry habitat among wild soybeans from habitats with different water availability [58]. On the other hand, comparison of midday paraheliotropism between two Japanese soybean cultivars exhibited that a major cultivar (Fukuyutaka) with stable seed yield had more vertical midday leaf angle than that of a local cultivar (Misato-zairai) with unstable seed yield and not erect leaves (**Figure 6** [61]).

Leaf angle is composed of the inclination angle of the petiole and the angle between the rachis and leaflets. The latter changes more largely than the former during the daytime (**Figure 6**, [61]), which is associated with paraheliotropism deeply. Meanwhile, there is also cultivar difference in the leaf orientation angle without paraheliotropic leaf movement; for example, this angle of Fukuyutaka is often higher than that of Misato-zairai. However, the correlation between the inclination angle and water content in petiole, which controls leaf orientation angle without paraheliotropism through the turgor change in the pulvinus, is similar in these two cultivars (**Figure 7**) [60]. So, the leaf orientation angle without

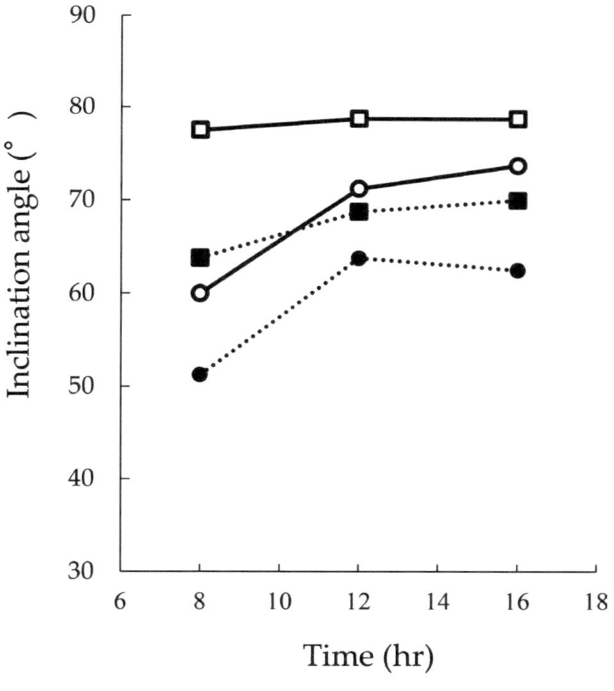

Figure 6.
Daily changes in inclination angles of leaflet and petiole in Fukuyutaka (open) and Misato-zairai (close) at beginning bloom (R1). Circle and square symbols represent leaflet and petiole, respectively. The inclination angles of the leaflet and petiole are meant to be the angle between the horizontal plane and the lines from the primary (stem-petiole) pulvinus to the tip of the central leaflet (leaflet) or to the secondary pulvinus (petiole) [61].

paraheliotropic leaf movement would depend on plant water condition through soil water condition and water transport capacity of soybean plants.

3.4 Leaf photosynthesis

Leaf photosynthetic rate means the apparent rate of photosynthetic CO_2 uptake per unit leaf area and is often expressed 'photosynthetic rate' or 'CO_2 exchange rate (CER)'. It is obvious that leaf photosynthesis is an important factor for determining the grain yield through canopy photosynthesis, and high yielding varieties of rice plants often have high leaf photosynthesis and its proper response to top dressing [62].

In Northeast China, it is indicated that leaf photosynthesis is one of the most important genetic contributors to the yield gain through the more plant biomass accumulation during the past 56 years [63]. Similar trends were found in North America [64–66]. For example, the experiment with new and old Canadian cultivars indicated that a 0.52% per yr. increases in leaf photosynthesis and this is a level very similar to the annual yield increase shown by these cultivars [67]. However, there are also conflicting reports on the relationship between leaf photosynthesis and yield; some researchers report only low to moderate correlations [8, 67], and others found no correlation [11, 68, 69]. The association between photosynthesis and yield is likely to be strong during seed filling period (R5–R7). Soybean leaf photosynthesis begins to decline around this period [70, 71], and this decline includes two types: slow and rapid [39]. The conflicting reports on the relationship between leaf photosynthesis and yield may be due to the phenological stage at which the measurements were taken [5]. Leaf photosynthesis is composed of mesophyll photosynthetic capacity and stomatal opening; the latter factor controls leaf photosynthesis through the CO_2 influx into the leaves. Stomatal opening also has another role in controlling H_2O efflux from the leaves; this varies in response to various environmental conditions to avoid excess transpiration and, as a result, leaf water shortage. Therefore, leaf photosynthesis is sensitive to leaf water potential (indicator of leaf water status); leaf photosynthesis of crop plants usually begins to drops in response to the decline in leaf water potential through the decline in stomatal

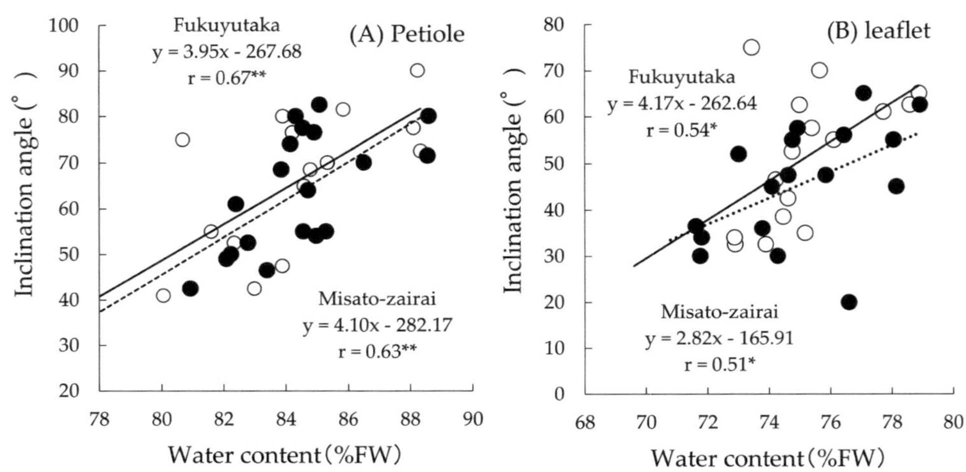

Figure 7.
*Variation in inclination angles as a function of water content of petiole (A) and leaflet (B) in Fukuyutaka (open) and Misato-zairai (close) at beginning bloom (R1) [60]. r is correlation coefficient. *, **: significant at 5% and 1%, respectively.*

conductance (indicator of stomatal opening) [72]. However, the sensitivity of soybean leaves to soil drought is lower; stomatal conductance (indicator of leaf water status) [73] and, as a result, leaf photosynthesis [74] are relatively unaffected until leaf water potential drops below −1.1 MPa and then drops dramatically [73]. These affect strongly the recovery from severe soil drought stress: leaf photosynthesis generally decreases with the decline in stomatal conductance even under mild soil drought condition to avoid excessive water loss from the leaves and severe damage to mesophyll photosynthetic capacity [72]. So, improvement of soil water condition (i.e. precipitation) can cause the recovery of leaf photosynthesis quickly thanks to the maintenance of mesophyll photosynthetic capacity. However, soybean leaves keep high stomatal conductance under relatively severe soil drought condition [73], and this is easy to result in severe damage to mesophyll photosynthetic capacity. In addition, soybean leaves can also keep leaf greenness under severe soil drought condition [52]. So, the symptom of drought stress in soybean leaves is not clear, and the recovery of leaf photosynthesis is impossible if soybean leaves are drooped by severe soil drought. Similarly, soybean leaf senescence can occur without a concomitant loss of leaf greenness [5]. It is too difficult to measure the net leaf photosynthetic capacity and to evaluate the relationship between leaf photosynthesis and yield accurately.

4. Yield and yield components

The increase in the potential capacity of photoassimilate supply (source) is an important matter for high crop yield; however, high yield results not only from the increase in source alone but also from the potential capacity of photoassimilate accepter (sink). The sink accepts and consumes the photoassimilate for its own growth and maintenance. Soybean major sink is the economically important harvest components (seed); soybean seed yield can be expressed as the function of yield components as follows:

$$\text{Yield} = \text{Pod number} \times \text{Seed number per pod} \times \text{Seed size}$$

Seed size (g per seed) is also expressed as 100 seed weight. Pod number (no. m^{-2}) can be separated into plant density (hill number, plant m^{-2}) and pod number per plant (no. $plant^{-1}$); plant density is relatively easy to control among yield components through sowing. Meanwhile, pod number can be also separated into pod per reproductive node (no.), reproductive node number per area (no. m^{-2}), percent reproductive nodes (%) and node number per area (no. m^{-2}). Many researchers examine the determinant period of seed yield in relation to the manipulation of light interception by shading, defoliation and wide row spacing at various growth periods and indicate that light interception during the period between R1 and R6 affects seed yield strongly through the response of pod and seed number [75]. Sink capacity is like a process of photoassimilate accumulation in a container; high volume for accumulation can accept a large amount of photoassimilate, and this results in high seed yield. Similar to other crop plants, seed number plays a determinant role in sink capacity, and this increase is often associated with high seed yield (**Figure 2B**) [75]. The period between R1 and R6 determines two numbers; pod number is determined critically by light interception during the period between V5 and R3 [39] and seed number is during R3–R6 [76]. Although pod number does not control seed yield as strongly as seed number (**Figure 2A, B**), the occurrence of seed number depends on pod number and reproductive closely (**Figure 2C**). High canopy photosynthesis during the period between R1 and R6 would affect soybean seed yield with the increase in pod number and seed number.

Sink capacity is also associated with source in a volumatic flow of photoassimilate, phloem loading and unloading. For example, the removal of wheat ear at grain filling period reduced about 50% flag leaf photosynthesis within 3–15 h, and the outflow of ^{14}C-labeled assimilates from the flag leaves (indicator of loading) also reduced remarkably [77]. Sink activity such as formation of flower, pod and seed is sensitive to environmental stress, and this decline decreases leaf photosynthesis through the restriction of phloem loading. Maintenance of sink activity contributes to high yield not only by the increases in pod number and seed number but also by activating leaf photosynthesis. However, another major sink, *Rhizobium japonicum* in the root nodules of soybean plants, is more sensitive to environmental condition than the host plants. For example, respiration of root nodules decreases below −0.4 MPa of root nodule water potential [78] even though leaf photosynthesis is kept until −1.1 MPa of leaf water potential. Root nodules consume much photoassimilates to fix atmospheric nitrogen into ammonium and subsequently ureides for long-range transport [79]. The major source of nitrogen accumulation is atmospheric nitrogen fixation by root nodule [80]; 100% (Brazil) [81] or 40–50% (Midwestern United States) [82] of nitrogen need depends on biological nitrogen fixation. The decline in root nodule activity would decrease not only canopy photosynthesis and seed filling directly by nitrogen deficiency but also leaf photosynthesis indirectly through the restriction of phloem loading and unloading, and this may be associated with the complex of soybean seed production.

5. Conclusion

As mentioned above, the researches about the relationship between seed production, yield components and canopy photosynthesis have been conflicted, and there is no worldwide universal theory for high yield. Even in Japan the strategy for high yielding is different; high biomass is useful for high yield in north Japan, and early sowing is tried to get high biomass [83]. However, high biomass often causes lodging and results in low yield in west Japan [84]; partial defoliation of above ground before blooming is recommended to avoid high biomass in Aichi Prefectures [84]. In the other major crops such as rice, maize and wheat, semidwarf gene plays an important role in high yield, and this introduction by breeding contributes to a dramatical increase of world grain yield [31, 85]. Dwarfing is also useful for soybean seed production; however, this factor alone cannot contribute to high yield. Soybean plants itself have acclimated flexibly to various environmental conditions, and the soybean cultivation countries increased about two times from 1961 to 2016 [86]. Realization of worldwide cultivation may be associated with the construction of independent growth and seed production proper to the area and its climate; therefore, it would be necessary for high soybean yield not only to expect the worldwide universal theory for high yield but also to build the own theory for each area in consideration of growth features and climate.

Author details

Kiyoshi Nagasuga
Graduate School of Bioresources, Mie University, Tsu, Mie, Japan

*Address all correspondence to: nkiyoshi@bio.mie-u.ac.jp

References

[1] MAFF Website 2018. Available from: http://www.maff.go.jp/j/seisan/ryutu/daizu/d_data/attach/pdf/index-24.pdf

[2] Kokubunm M. Genetic and cultural improvement of soybean for waterlogged conditions in Asia. Field Crops Research. 2012;**152**:3-7

[3] Cui SY, Yu DY. Estimates of relative contribution of biomass, harvest index and yield components to soybean yield improvements in China. Plant Breeding. 2005;**124**:473-476

[4] MAFF Website 2018. Available from: http://www.maff.go.jp/j/seisan/ryutu/daizu/d_data/attach/pdf/index-11.pdf

[5] Kumudini S. Trial and tribulations: A review of the role of assimilate supply in soybean genetic yield improvement. Field Crops Research. 2002;**75**:211-222

[6] Cregan PB, Yaklich RW. Dry matter and nitrogen accumulation and partitioning in selected soybean genotypes of different derivation. Theoretical and Applied Genetics. 1986;**72**:782-786

[7] Shibles R, Sundgerg DN. Relation of leaf nitrogen content and other traits with seed yield of soybean. Plant Production Science. 1999;**1**:3-7

[8] Morrison MJ, Voldeng HD, Cober ER. Physiological changes from fifty-eight years of genetic improvement of short-season soybean cultivars in Canada. Agronomy Journal. 1999;**91**:685-689

[9] Kumudini S, Hume DJ, Chu G. Genetic improvement in short season soybean cultivars [Ph.D. Dissertation]. Guelph, Canada: University of Guelph; 2001

[10] Song FS, Agata W, Kawamitsu Y. Studies on dry matter and grain production of F1 hybrid Rice in China I Characteristics of dry matter production. Japanese Journal of Crop Science. 1990;**59**:19-28. In Japanese with English abstract

[11] Gay S, Egli DB, Reicosky DA. Physiological aspects of yield improvement in soybeans. Agronomy Journal. 1980;**72**:387-391

[12] Schapaugh WT Jr, Wilcox JR. Relationships between harvest indices and other plant characteristics in soybeans. Crop Science. 1980;**20**:529-533

[13] Frederick JW, Woolley JT, Hesketh JD, Peters DB. Seed yield and agronomic traits of old and modern soybean cultivars under irrigation and soil water-deficit. Field Crops Research. 1991;**27**:71-82

[14] Shiraiwa T, Hashikawa U. Accumulation and partitioning of nitrogen during seed filling in old and modern soybean cultivars in relation to seed production. Japanese Journal of Crop Science. 1995;**64**:4-759

[15] Hirasawa T. Crop growth and physiology. In: General Topics of Crop Science. Tokyo: Asakura Publishing Co., Ltd; 1999. pp. 75-116. Translation of Japanese into English by the author

[16] Imsande J. Rapid dinitrogen fixation during soybean pod filling enhances net photosynthetic output and seed yield: A new perspective. Agronomy Journal. 1989;**81**:549-556

[17] Fageria NK, Baligar VC, Clark RB. Photosynthesis and crop yield. In: Physiology of Crop Production. New York: Food Products Press; 2006. pp. 95-116

[18] Loomis RS, connor DJ. Crop ecology. In: Productivity and Management in Agricultural System. Cambridge, U.K.:

Cambridge University Press; 1992. pp. 32-39

[19] Dry Mass, Biology Online Dictionary. 2018. Available from: https://www.biology-online.org/dictionary/Dry_mass

[20] Fehr WR, Caviness CE. Stage of Soybean Development, Special report No. 80. Ames, USA: Cooperative Extension Service, Agric. And Home Econ. Stn. Iowa State University; 1977

[21] Nagasuga K, Uchida S, Kaji H, Hayakawa Y, Nose S, Umezaki T. Seed production, dry matter production, and light intercepting characteristics of soybean cultivar 'Misato-zairai'. Japanese Journal of Crop Science. 2011;**80**:326-332. In Japanese with English abstract

[22] Board JE, Tan Q. Assimilatory capacity effects on soybean yield components and pod formation. Crop Science. 1995;**35**:846-851

[23] Egli DB, Crafts-Brandner SJ. Soybean. In: Photoassimilate Distribution in Plants and Crops—Source Sink Relationships. New York: Marcel Dekker Inc; 1996. pp. 595-623

[24] Van Schaik PH, Probes AH. Effects of some environmental factors on flower production and reproductive efficiency in soybeans. Agronomy Journal. 1958;**50**:192-197

[25] Hansen WR, Shibles R. Seasonal log of the flowering and podding activity of field-grown soybeans. Agronomy Journal. 1976;**70**:47-50

[26] Jiang H, Egli DB. Shade induced changes in flower and pod number and flower and fruit abscission in soybean. Agronomy Journal. 1992;**85**:221-225

[27] Ohnishi S, Miyoshi T, Shirai S. Low temperature stress at different flower developmental stages affects pollen development, pollination, and pod set in soybean. Environmental and Experimental Botany. 2010;**69**:56-62

[28] Westgate ME, Peterson CM. Flower and pod development in water-deficient soybean (*Glycine max* L. Merr.). Journal of Experimental Botany. 1993;**44**:109-117

[29] Jiang H, Egli DB. Shade induced changes in flower and pod number and flower and fruit abscission in soybean. Agronomy Journal. 1993;**85**:221-225

[30] Saitoh K, Nishimura K, Kuroda T. Characteristics of flowering and pod set in wild and cultivated types of soybean. Plant Production Science. 2004;**2**:172-177

[31] Ishii R. Leaf/canopy photosynthesis and crop productivity. In: Photosynthesis: A Comprehensive Treatise. Cambridge, U.K.: Cambridge University press; 1998. pp. 215-225

[32] Egli DB. Cultivar maturity and potential yield of soybean. Field Crops Research. 1993;**32**:147-158

[33] Shibles RM, Weber CR. Leaf area, solar radiation interception and dry matter production of soybeans. Crop Science. 1965;**5**:575-577

[34] Weber CR, Shibles RM, Byth DE. Effect of plant population and row spacing on soybean development and production. Agronomy Journal. 1966;**58**:99-102

[35] Hardman LL, Brun WA. Effect of atmospheric carbon dioxide enrichment at different developmental stages on growth and yield components of soybeans. Crop Science. 1971;**11**:886-888

[36] Board JE, Harville BG. Soybean yield component responses to a light interception gradient during the reproductive period. Crop Science. 1993;**33**:772-777

[37] Board JE, Zhang W, Harville BG. Yield rankings for soybean cultivars grown in narrow and wide rows with late planting dates. Agronomy Journal. 1996;**88**:240-245

[38] Board JE, Wier AT, Boethel DJ. Source strength influence on soybean yield formation during early and late reproductive development. Crop Science. 1995;**35**:1104-1110

[39] Mathew JP, Herbert SJ, Zhang S, Rautenkranz AF, Litchfield GV. Differential response of soybean yield components to the timing of light enrichment. Agronomy Journal. 2001;**92**:1156-1161

[40] Hayati R, Egli DB, Crafts-Brandner SJ. Carbon and nitrogen supply during seed filling and leaf senescence in soybean. Crop Science. 1995;**35**:1063-1069

[41] Nagasuga K, Kadowaki M, Uchida S, Kaji H, Fukunaga A, Umezaki T. Effects of water condition on soybean (*Glycine max* L.) plant growth after flowering. Environmental Control in Biology. 2014;**52**:221-225

[42] Nagasuga K, Hayakawa Y, Kadowaki M, Fukunaga A, Nose S, Umezaki T. Comparison of light-intercepting characteristics between soybean (*Glycine max*) cultivar 'Misato-zairai' and 'Fukuyutaka'. Environmental Control in Biology. 2013a;**51**:85-88

[43] Nagasuga K, Kanai R, Kadowaki M, Sasayama N, Umezaki T. Characteristics of dry matter production in soybean cultivar 'Misato-zairai' through its response to soil water condition. Abstract of the 243th Meeting of the CSSJ. 2017;48. In Japanese

[44] Nagasuga K, Kanai R, Kadowaki M, Sasayama N, Umezaki T. Examination of seed production characteristics in soybean cultivar 'Misato-zairai' through its response to soil water condition. Abstract of the 243th Meeting of the CSSJ. 2017;49. In Japanese

[45] Nakayama N, Shimada S, Takahashi M, Kim YH, Arihara J. Effects of water absorbing rate of seed on flooding injury in soybean. Japanese Journal of Crop Science. 2005;**74**:325-329. In Japanese with English abstract

[46] Tian XH, Nakamura T, Kokubun M. The role of seed structure and oxygen responsiveness in pre-germination flooding tolerance of soybean cultivars. Plant Production Science. 2005;**8**:157-165

[47] Masuda Y, Onnaka N. Effect of planting density on growth and forage yield of soiling groups I. On soiling soy bean. Japanese Journal of Grassland Science. 1966;**11**:114-122. In Japanese with English abstract

[48] Inoue H, Ito T, Saigusa M. Effects of planting density and application rate of nitrogen fertilizer on lodging and yield of dent corn tillage system. Japanese Journal of Grassland Science. 2000;**46**:249-253. In Japanese with English abstract

[49] Kobayashi R, Sato K, Hattori I, Konishi T. Optimum planting density and fertilizer application rate of forage rice under ratoon cropping in the central Kyushu region. Japanese Journal of Grassland Science. 2007;**53**:208-214. In Japanese with English abstract

[50] PBIO*3110. Crop Physiology. 2018. Available from: http://greenlab.cirad.fr/GLUVED/html/P1_Prelim/EPhysio/lec02_08.pdf

[51] Shibles R, Secor J, Ford DM. Carbon assimilation and metabolism. In: Wilcox JR, editor. Soybeans: Improvement, Production, and Uses. Madison, WI, USA: American Society of Agronomy-Crop Science Society of America-Soil Science Society of America, Amer. Soc. Of Agron; 1987. pp. 535-588

[52] Nagasuga K, Fukunaga A, Higashi C, Umezaki T. The Response of vegetative growth to soil water condition in native soybean cultivar 'Misato-zairai'. Japanese Journal of Crop Science. 2016;**85**:138-143. In Japanese with English abstract

[53] Nagasuga K, Fukunaga A, Higashi C, Umezaki T. The response of dry matter production to soil water condition in soybean cultivar 'Misato-zairai' and 'Fukuyutaka' grown in different air temperature conditions. Abstract of the 240th Meeting of the CSSJ. 2015;52. In Japanese

[54] Moldenhauer KAK, Gibbons JH. Rice morphology and development. In: Rice: Origin, History, Technology, and Production. Hoboken, N.J: John Wiley & Sons, Inc; 2003. pp. 103-127

[55] Soybean Growth and Development. University of Wisconsin Website. 2018. Available from: http://corn.agronomy.wisc.edu/Default.aspx

[56] Wells R, Schulze FH, Ashley DA, Boerma HR, Brown RH. Cultivar differences in canopy apparent photosynthesis and their relationship to seed yield in soybean. Crop Science. 1982;**22**:886-890

[57] Critchley C. Photoinhibition. In: Photosynthesis: A Comprehensive Treatise. Cambridge, U.K.: Cambridge University press; 1998. pp. 264-272

[58] Kao WY, Tsai TT. Tropic leaf movements, photosynthetic gas exchange, leaf^{13}C and chlorophyll a fluorescence of three soybean species in response to water availability. Plant, Cell & Environment. 1998;**21**:1055-1062

[59] Toriyama H. Observational and experimental studies of sensitive plants. I: The structure of parenchymatous cells of pulvinus. Cytologia. 1955;**18**:283-291

[60] Nagasuga K, Uchida S, Kaji H, Hayakawa Y, Kadowaki M, Fukunaga A, et al. Water condition controls inclination angles of leaflets and petioles of soybean (*Glycine max* L.). Environmental Control in Biology. 2013b;**51**:79-84

[61] Nagasuga K, Kaji H, Uchida S, Hayakawa Y, Nose S, Umezaki T. Comparison of paraheliotropic leaf movement between the soybean cultivars 'Misao-zairai' and Fukuyutaka. Annals of Field Research and Technology, Mie University. 2013;**11**:1-8. In Japanese with English abstract

[62] Sasaki H, Ishii R. Cultivar differences in leaf photosynthesis of rice bred in Japan. Photosynthesis Research. 1992;**32**:139-146

[63] Jian J, Xiaobing L, Guanghua W, Liang M, Zhongbao S, Xueli C, et al. Agronomic and physiological contributions to the yield improvement of soybean cultivars released from 1950 to 2006 in Northeast China. Field Crops Research. 2010;**115**:116-123

[64] Dornhoff GM, Shiebles RM. Varietal differences in net photosynthesis of soybean leaves. Crop Science. 1970;**10**:42-46

[65] Buttery BR, Buzzell RI, Findlay WI. Relationships among photosynthetic rate, bean yield, and other characters in held-grown cultivars of soybean. Canadian Journal of Plant Science. 1981;**61**:191-198

[66] Morrison MJ, Voldeng HD, Cober ER. Agronomic changes from 58 years of genetic improvement of short-season soybean cultivars in Canada. Agronomy Journal. 2000;**9**:780-784

[67] Buttery BR, Buzzell RI. Soybean leaf nitrogen in relation to photosynthetic rate and yield. Canadian Journal of Plant Science. 1988;**68**:793-795

[68] Ford DM, Shibles R, Green DE. Growth and yield of soybean

lines selected for divergent leaf photosynthetic ability. Crop Science. 1983;**23**:517-520

[69] Frederick JR, Aim DM, Hesketh JD. Leaf photosynthetic rates, stomatal resistances, and internal CO_2 concentrations of soybean cultivars under drought stress. Photosynthetica. 1989;**23**:575-584

[70] Sinclair TR. Leaf CER from post-flowering to senescence of field-grown soybean cultivars. Crop Science. 1980;**20**:196-200

[71] Lugg DG, Sinclair TR. Seasonal changes in photosynthesis of field-grown soybean leaflets: 2. Relation to nitrogen content. Photosynthetica. 1981;**15**:138-144

[72] Berkowitz GA. Water and salt stress. In: Photosynthesis: A Comprehensive Treatise. Cambridge, U.K.: Cambridge University Press; 1998. pp. 226-237

[73] Nielsen DC. Scheduling irrigations for soybeans with the crop water stress index (CWSI). Field Crops Research. 1990;**23**:103-116

[74] Boyer JS. Leaf enlargement and metabolic rates in corn, soybean, and sunflower at various leaf water potentials. Plant Physiology. 1970;**46**:233-235

[75] Board JE, Kumudini S, Omielan J, Prior E, Kahlon CS. Yield response of soybean to partial and total defoliation during the seed-filling period. Crop Science. 2010;**50**:703-712

[76] Egli DB. Soybean reproductive sink size and short-term reductions in photosynthesis during flowering and pod set. Crop Science. 2010;**50**:1971-1977

[77] King RW, Wardlaw IF, Evans LT. Effect of assimilate utilization on photosynthetic rate in wheat. Planta. 1967;**77**:261-276

[78] Pankhurst CE, Sprent JI. Effects of water stress on the respiratory and nitrogen-fixing activity of soybean root nodules. Journal of Experimental Botany. 1975;**26**:287-304

[79] Collier R, Tegeder M. Soybean ureide transporters play a critical role in nodule, development, function and nitrogen export. The Plant Journal. 2012;**72**:355-367

[80] Jin J, Liu X, Wang G, Liu J, Mi L, Chen X, et al. Leaf nitrogen status as a main contributor to yield improvement of soybean cultivars. Agronomy Journal. 2011;**103**:441-448

[81] Dobereiner J. Biological nitrogen fixation in the tropics–Social and economic contributions. Soil Biology and Biochemistry. 1997;**29**:771-774

[82] Ham GE. Interactions of *Glycine max* and *Rhizobium japonicum*. In: Advances in Legume Science. Kew: Royal Botanic Gardens; 1978. pp. 289-296

[83] Kumagai E. Effect of early sowing on growth and yield of determinate and indeterminate soybean (*Glycine max* (L.) Merr.) cultivars in a cool region of northern Japan. Journal of Agricultural Meteorology. 2018;**74**:18-28

[84] Hayashi M, Hamada Y, Tani T, Hiraiwa K. Development of pinching machine and effect of pinching for soybean. Research Bulletin of the Aichi Agricultural Research Center. 2008;**40**:93-97. In Japanese with English abstract

[85] Sakamoto T, Morinaka Y, Ohnishi T, Sunohara H, Fujioka S, Ueguchi-Tanaka M, et al. Erect leaves caused by brassinosteroid deficiency increase biomass production and grain yield in rice. Nature Biotechnology. 2006;**24**:105-109

[86] Crops (production), FAOSTAT. 2018. Available from: http://www.fao.org/faostat/en/#data/QC

Chapter 4

Seedling Shoot and Root Growth Responses among Soybean (*Glycine max*) Genotypes to Drought Stress

Obed J. Mwenye, Leon Van Rensburg, Angeline Van Biljon and Rouxlene Van der Merwe

Abstract

Drought stress is a major limiting factor in soybean production in South Africa. The development of soybean varieties with enhanced tolerance to soil waterlimited induced stress (WLIS) is one sustainable way to deal with drought. Root traits have shown strong potential for improvement of drought tolerance through breeding. The objectives of this study were to evaluate seedling shoot- and root growth responses under WLIS in order to study root morphology as a mechanism to cope with drought stress and to determine if there were genotypic differences in shoot- and root morphology between drought tolerant and -sensitive soybean genotypes. Seedlings of three drought tolerant and one sensitive genotype were subjected to soil WLIS in deep-root-pots for 21 days. Results suggested significant genotypic differences for shoot length, number of leaves, tap root length and root-to-shoot length ratio. Soybean tolerant genotypes were associated with moderate shoot biomass, deep rooting abilities and maintained a large root-to-shoot ratio under WLIS conditions. In contrast, the sensitive genotype was associated with a reduced root-to-shoot ratio and shallower root system. Soybean genotypes showed varying seedling root growth responses to soil WLIS, while shoot biomass characteristics were similar.

Keywords: root, drought, tolerance, selection, soybean

1. Introduction

Soybean (*Glycine max* L. Merrill) is the world's leading economic oilseed crop [1]. It is the largest source of vegetable oil and protein in the world [2] and a primary source of protein in the livestock feed industry in South Africa [3]. Soybean production in South Africa is limited by inadequate rainfall and soil water-limited-induced-stress (WLIS) which characterises most production regions [3, 4], consequently affecting the livestock industry and food security of the country [5].

Direct selection for morphological and physiological parameters related to soil WLIS tolerance has been used in developing drought tolerant soybean cultivars [6]. Drought tolerant mechanisms in soybean have been closely associated with the rooting system and/or rooting pattern [7, 8]. The response of root traits to the immediate effects of soil WLIS determines the soybean plants' defence to drought [7, 9]. This is because the root system is responsible for exploring and acquisition of all the water the plant requires from the soil [10]. Plants sense water shortage around their roots and respond instantaneously by sending chemical signals to shoots to initiate various adaptive responses to soil WLIS [11].

Genetic variation for consultative ability and phenotypic plasticity in root growth patterns have been reported in soybean [12, 13]. Although root traits have shown strong potential in breeding for drought tolerance in soybean [6], the laborious and difficult procedure involved to phenotype roots without breaking the tap root is a big setback [14]. However, a "deep-pot" root screening method, which facilitates removal of the root system with minimal damage and evaluation of the soybean root architecture, has been developed [9]. This screening method facilitates root evaluation efficiently at seedling stage. The present study was conducted with the aim of discriminating drought tolerant and –sensitive genotypes through root- and shoot morphology. Specific objectives were to determine the responses of seedling shoot- and root length, dry matter and root length density among soybean genotypes grown under soil WLIS using the "deep-pot" method. Study using the same four PANNAR SEED® (Pty) LTD soybean cultivars, under similar experimental conditions as at the University of Free State, South Africa has not been done before.

2. Materials and methods

2.1 Plant material and trial site

Glasshouse trials were conducted in duplicate (trial 1 from 06/09/2014 to 07/10/2014 and trial 2 from 06/11/2014 to 08/12/2014) at the Department of Soil, Crop and Climate Sciences of the University of the Free State, South Africa in the summer season. The plant material used included three drought tolerant genotypes (BL1, BL4 and CV2) and one susceptible genotype (CV3) obtained from the PANNAR SEED® (PTY) LTD breeding programme (**Table 1**). The plant materials in the study are grouped in the medium quick (5.7), medium (6.7) and medium late (7.0–7.4) maturity groups (**Table 1**) [15].

Glasshouse trials were conducted in deep transparent polythene pots (10 cm diameter and 110 cm length) according to [9] with modifications. Polythene pots were placed in polyvinyl chloride (PVC) tubes (10 cm diameter and 10 cm length) for support and the pots were filled with soil, leaving a 5 cm space at the top (**Figure 1**). Soil used was an aridic ustothent soil (Bainsvlei 2300) with a reddish brown colour

Genotype	Growth habit	Maturity group	Drought sensitivity
BL1	Determinate	6.7	Tolerant
BL4	Indeterminate	7.4	Tolerant
CV2	Indeterminate	7.0	Tolerant
CV3	Determinate	5.7	Sensitive

Maturity group (length of the growing season of a cultivar, mostly determined by day length and temperatures): 5.7 = medium quick, 6.7 = medium, 7.0 &7.4 = medium late.

Table 1.
Soybean genotypes used in the study and their levels of drought sensitivity.

Figure 1.
"Deep-pot" system used for screening shoot- and root morphology of four soybean seedling genotypes grown under water-limited-induced-stress conditions for 21 days from sowing.

and a fine sandy texture, and contained 8–14% clay and 2–4% silt [16]. Pots were fully saturated with water and left to drain. After seven days, three seeds from each genotype were sown in the pots.

2.2 Experimental layout and data collection

The pots were laid out in a randomised complete block design with six replications. After germination, seedlings were thinned to one per pot. For the first 12 days after sowing, plants were watered daily with 100 ml of water to ensure strong seedlings, thereafter soil WLIS was induced by irrigating 100 ml every third day. Twenty-one days after sowing (**Figure 1**), shoot length (SL) (cm), was measured from the soil contact point to the tip of the plant and number of trifoliate leaves was counted. Shoot dry weight (biomass) (g) was recorded after drying the shoots in the glasshouse at room temperature for 72 hours.

For root measurements, the polythene pots were carefully taken from the PVC tubes and cut longitudinally in order to safely isolate the whole root system from the soil. Tap root length (TRL) was measured on intact roots from the soil contact point to the tip. Total root dry weight (root biomass) (g) was recorded after separating the roots from the soil by washing the samples on a 0.5 mm sieve. The root samples were then dried at room temperature, weighed and root biomass determined. After recording tap root length, shoot and root biomass, all roots including tap root and primary roots were cut into 10 cm sections, corresponding to the actual depths from the tubes. For root length density, root samples in the 10 cm sections were counted using a modified infra-red root counter [17] and were converted into root length per soil layer using a standard curve generated from standard samples and actual root length. Mass of the total root per layer was used to calculate the root weight density per layer.

2.3 Data analysis

Data was statistically analysed using GenStat Release 18 statistical package [18]. A combined analysis of variance (ANOVA) (**Table 2**) was conducted for the duplicated trials on the data collected to partition the different sources of variation. Means were separated using least significant difference (LSD) test (**Table 2**).

Source of variation	Shoot length (cm)	No. of trifoliate leaves	Shoot dry weight (g)	Tap root length (cm)	Root dry weight (g)	TRL/SL	RDW/SDW
Replication	2.38	1.02	0.07	70.70	0.03	0.39	0.01
Trial (T)	1.02	1.02	1.56**	78.70	0.51**	0.22	0.01
Genotype (G)	40.34**	7.74**	0.11	1552.60**	0.02	4.76*	0.02
G×T	12.56	1.74	0.05	53.50	0.04	1.53	0.02
Residual	10.80	1.07	0.13	170.50	0.02	1.28	0.01
CV%	18.80	26.80	23.70	16.30	23.60	24.00	24.70

**, **, significant at $P < 0.05$, $P < 0.01$, respectively. CV = coefficient of variation, TRL/SL = tap root length to shoot length ratio, RDW/SDW = root dry weight to shoot dry weight ratio, G×T = genotype by trial interaction.*

Table 2.
Combined analysis of variance showing mean square values for shoot- and root traits of soybean seedlings under water-limited-induced stress at 21 days after sowing.

3. Results

3.1 Response of shoot and root traits

Significant genotype differences were observed for shoot length ($p < 0.01$), number of leaves ($p < 0.01$), tap root length ($p < 0.01$) and root-to-shoot length ratio (TRL/SL) ($p < 0.05$) (**Table 2**). Trial effects were significant for shoot and root dry weight ($p < 0.01$). No significant genotype by treatment (G×T) interaction effects were observed for any of the traits measured (**Table 2**).

This suggested that genotypes responded the same between the two trials for all traits studied. Tolerant genotypes BL1 and BL4 grouped together for shoot length (both large), while the other tolerant genotype CV2 grouped with the sensitive genotype CV3 (**Table 3**). This, together with shoot dry weight (non-significant genotype effects), indicated that the sensitive genotype CV3 generally showed a good shoot biomass and was not different from the tolerant genotypes, especially BL4 and CV2 (**Table 3**).

Source of variation	Shoot length (cm)	No. of trifoliate leaves	Shoot dry weight (g)	Tap root length (cm)	Root dry weight (g)	TRL/SL	RDW/SDW
BL1	19.58^{a}	4.42^{a}	1.55	80.10^{b}	0.60	$4.22b^{c}$	0.40
BL4	18.12^{ab}	3.08^{b}	1.59	95.50^{a}	0.69	5.37^{a}	0.45
CV2	15.25^{c}	3.25^{b}	1.37	75.30^{bc}	0.67	5.13^{ab}	0.49
CV3*	16.96^{bc}	4.67^{a}	1.50	68.80^{c}	0.67	4.13^{c}	0.45
Mean	*17.33*	*3.85*	*1.50*	*79.90*	*0.66*	*4.71*	*0.45*
$LSD_{0.05}$	2.72	0.86	0.29	10.82	0.13	0.94	0.09

**Drought sensitive genotype. TRL/SL = tap root length to shoot length ratio, RDW/SDW = root dry weight to shoot dry weight ratio, LSD = least significant difference. Means followed by the same letter of the alphabet in the column are not significantly different.*

Table 3.
Combined mean values for shoot- and root characteristics of soybean seedlings under water-limited-induced stress conditions at 21 days after sowing.

Figure 2.
Root lengths of four soybean seedling genotypes with CV3 as the sensitive genotype under soil water-limited-induced-stress conditions at 21 days after sowing.

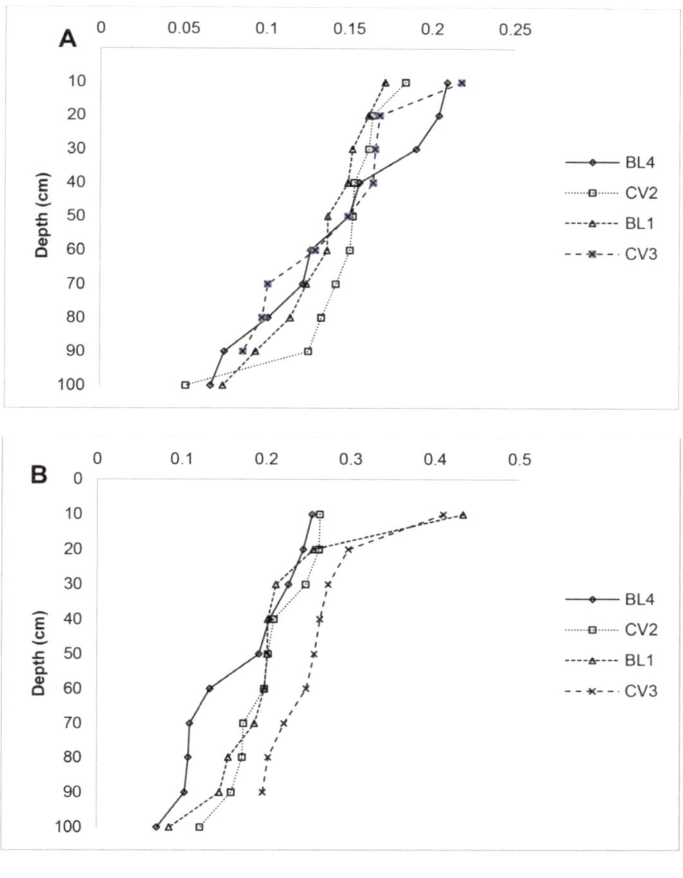

Figure 3.
Root length density (cm cm^{-3}) distribution with depth for four soybean seedlings grown under soil water-limited-induced stress conditions for duplicate trials: (A) trial 1 and (B) trial 2.

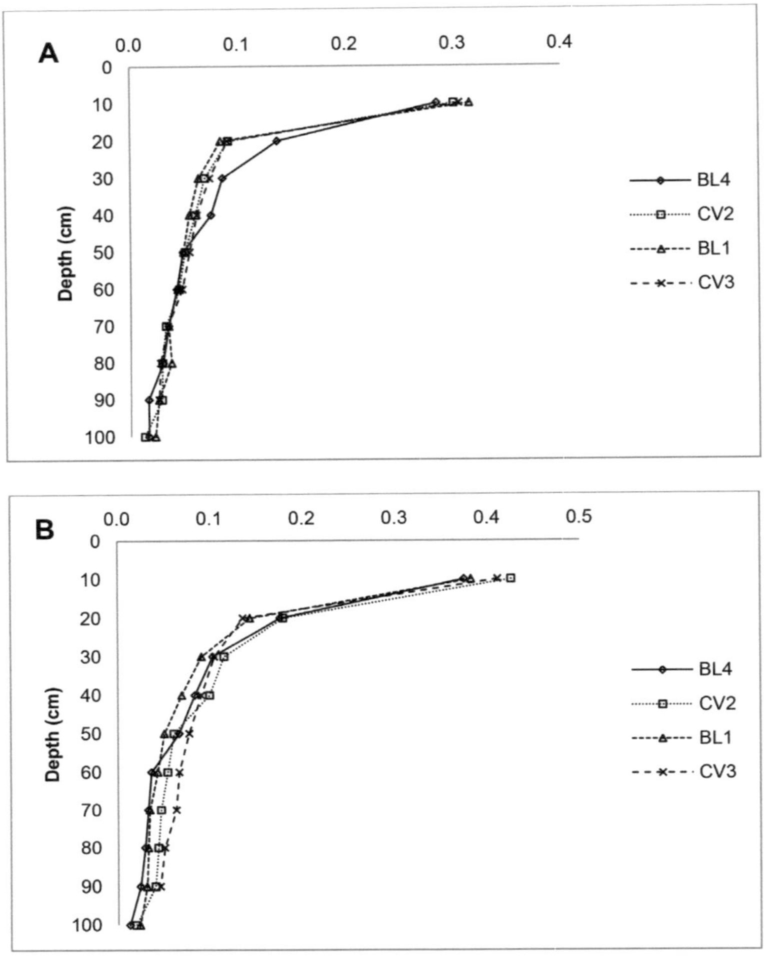

Figure 4.
Root weight density (mg cm^{-3}) distribution with depth for four soybean seedlings grown under soil water-limited-induced stress conditions for duplicate trials: (A) trial 1 and (B) trial 2.

The three tolerant genotypes (BL1, BL4 and CV2) ranked in the top three positions for both tap root length and TRL/SL ratio, while the drought sensitive genotype (CV3) ranked last (**Table 3**). This suggested that the drought sensitive genotype developed a shorter tap root (as indicated in **Figure 2**) or had a reduced root elongation rate compared to the tolerant genotypes due to the soil WLIS. In **Figure 2** it is evident that the three tolerant genotypes had longer root systems than the sensitive genotype (CV3).

The root architecture system under soil WLIS was investigated by determining the distribution of the root length density (cm cm^{-3}) and dry matter (root weight density) (mg cm^{-3}) through the soil profile (**Figures 3** and **4**). Results showed that the root systems of all four cultivars were mostly contained in the top 0–40 cm soil profile (**Figures 3** and **4**). There were no marked differences in root distribution among the tolerant genotypes (BL1, BL4 and CV2) (**Figures 3** and **4**). The tolerant genotypes' roots explored the whole soil profile to the deepest part (**Figures 3** and **4**).
On the other hand, it was evident that the sensitive genotype (CV3) had the same root distribution (root length density and dry matter) like the tolerant genotypes in the top 0–90 cm soil profile (**Figures 3** and **4**). However, the sensitive genotype's roots could not explore the deeper profile beyond the 90 cm depth mark (**Figures 3** and **4**).

This suggested that the tolerant genotypes, unlike the sensitive one, could endure the soil WLIS by exploring moisture trapped in the deep soil profiles.

4. Discussion

Drought stress tended to increase biomass partitioning to the roots. This was demonstrated by the increase in the tap root length to shoot length ratio (TRL/SL) especially among the tolerant soybean genotypes (**Figures 3** and **4**). Similar trends were reported by [19] using a deep-rooted pot system with 24 soybean cultivars including exotic plant introductions (PI) lines in where it was suggested that TRL/SL determines the effective proportion of the roots supporting the above ground shoots. This is significant for survival under soil WLIS conditions. The large tap root length to shoot length ratio observed for the drought tolerant genotypes might have translated to the ability of the tolerant cultivars to access and get more edaphic resources per unit of the above ground shoot [20, 21]. A large TRL/SL suggests a large root surface area per unit of shoot length and this significantly increases the capacity of the tolerant genotypes to efficiently absorb and utilise soil water per unit dry matter [10]. The drought sensitive soybean genotype indicated to have a lower ability to penetrate the deeper soil (**Table 3**).

The seedling genotypic variation observed between drought tolerant and –sensitive genotypes for root and shoot traits, in response to soil WLIS (**Table 3**) agrees with results of [7, 11, 22, 23]. Drought tolerant soybean genotypes were associated with a deep-rooting phenotype and a large root-to-shoot ratio (length and/or mass), unlike the drought sensitive soybean genotypes (**Table 3**). Thus, drought tolerant genotypes tend to demonstrate a phenomenon called "balanced growth", whereby plants respond to drought by stimulating or maintaining root growth while effectively reducing shoot growth [24]. Increases in root versus shoot growth under drought conditions enhance the genotype's drought coping ability due to increased root-to-leaf surface ratio, continued production of new root tips, and enhancement of plant capacity for acquiring water to support existing shoots [14]. The observed shift in allometry under drought stress and the ability of the tolerant genotypes to grow roots according to the distribution of available soil water thus increase the productivity of tolerant genotypes under soil WLIS conditions. This has been attributed to the action of ethylene and abscisic acid in the tolerant soybean genotypes [25, 26] and is triggered by soil WLIS conditions.

5. Conclusions and recommendations

Soybean genotypes showed varying seedling shoot- and root morphology in response to soil WLIS conditions. The tolerant and sensitive genotypes did not differ much for shoot- and root biomass but significant differences were observed for shoot length, tap root length and tap root length to shoot length ratio. Drought tolerant genotypes showed deep rooting ability and larger root-to-shoot ratios compared to the drought sensitive genotype. Consequently, the drought tolerant genotypes increased biomass partitioning to the roots as a coping mechanism to soil WLIS. The "deep-pot" system was effectively used to phenotype the roots of soybean seedlings in a non-destructive manner. The system showed potential to differentiate between drought tolerant and -sensitive genotypes at seedling growth stage. However, in future studies more genotypes and more root traits need to be analysed in order to efficiently classify genotypes as drought tolerant or drought sensitive.

Acknowledgements

National Research Foundation (NRF), South Africa, for research funding.

The University of the Free State, Strategic Academic Cluster: Technologies for sustainable crop industries in semi-arid regions programme for the study scholarship bursary and for research funding.

PANNAR® Seed (PTY) LTD breeding programme, in Greytown, South Africa for the plant materials used in the study.

Author details

Obed J. Mwenye[1,3]*, Leon Van Rensburg[2], Angeline Van Biljon[1] and Rouxlene Van der Merwe[1]

1 Department of Plant Sciences, University of the Free State, Bloemfontein, South Africa

2 Department of Soil, Crop and Climate Sciences, University of the Free State, South Africa

3 Department of Agricultural Research Services, Bvumbwe Research Station, Limbe, Malawi

*Address all correspondence to: omwenye@yahoo.co.uk

References

[1] Tidke SA, Ramakrishna D, Kiran S, Kosturkova S, Ravishankar GA. Nutraceutical potential of soybean: Review. Asian Journal of Clinical Nutrition. 2015;**7**:22-32

[2] Corner T, Paschal EH, Barbero A, Johnson F. The challenges and potential for future agronomic traits in soybeans. AgBioforum. 2004;**7**:47-50

[3] Dlamini TS, Tsabalala P, Mutengwa T. Soybeans production in South Africa. Oilseeds and fats, crops and lipids: Oil crops and supply chain in Africa. 2014;**21**:1-11. DOI: 10.1051/ocl/2013047

[4] Kobraei S, Shamsi K, Rasekhi B. Soybean production under water deficit conditions. Annals of Biological Research. 2011;**2**:423-434

[5] Ngaka MJ. Drought preparedness, impact and response. A case of the Eastern Cape and Free State provinces of South Africa. Journal of Disaster Risk Studies. 2012;**4**:1-10

[6] Manavalan LP, Guttikonda SK, Tran LS, Nguyen HT. Physiological and molecular approaches to improve drought resistance in soybean. Plant Cell Physiology. 2009;**50**:1260-1276

[7] Fenta BA, Beebe SE, Kunert KJ, Burridge J, Barlow KM, Lynch JP, et al. Field phenotyping of soybean roots for drought stress tolerance. Agronomy. 2014;**4**:418-435

[8] Nguyen BAT, Nguyen QT, Hoang XL, Thao NP, Tran LSP. Evaluation of drought tolerance of the Vietnamese soybean cultivars provides potential resources for soybean production and genetic engineering. BioMed Research International. 2014;**2014**:809736. DOI: 10.1155/2014/809736 [Accessed: 27/07/2017]

[9] Manavalan LP, Guttikonda SK, Nguyen VT, Shannon JG, Nguyen HT. Evaluation of diverse soybean germplasm for root growth and architecture. Plant and Soil. 2010;**330**:503-514

[10] Malamy J. Intrinsic and environmental response pathways that regulate root system architecture. Plant Cell and Environment. 2005;**28**:67-77

[11] Hossain MM, Liu X, Qi X, Lam HM, Zhang J. Differences between soybean genotypes in physiological response to sequential soil drying and rewetting. The Crop Journal. 2014;**2**:366-380

[12] Manavalan LP, Musket T, Nguyen HT. Natural genetic variation for root traits among diversity lines of maize (*Zea mays* L.). Maydica. 2011;**56**:1707. (Advance Access publication 2011)

[13] Mastuo N, Takahashi M, Fukami K, Tsuchiya S, Tasaka K. Root growth of two soybean [*Glycine max* (L.) Merr.] cultivars grown under different groundwater level conditions. Plant Production Science. 2013;**16**:374-382

[14] Cosmas LH, Becker SR, Cruz VMV, Byrne PF, Dierg DA. Root traits contributing to plant conductivity under drought. Frontiers in Plant Science. 2013;**4**:1-19

[15] Soybeans production guide. Pannar Seed (Pty). Soybeans Production Guide for South Africa. 2006. http://www.pannar.com/assets/documents/soybeans.pdf. [Accessed: 20/08/2018]

[16] Soil Classification Working Group. Soil classification, a taxonomic system for South Africa. Memoirs on the Agricultural Natural Resources of South Africa No. 15. Pretoria: Department of Agricultural Development; 1991. pp. 257

[17] Rowse HR, Phillips DA. An instrument for estimating the total length of root in sample. Journal of Applied Ecology. 1974;**11**:309-314

[18] VSN Internatiomal. Gensta® 18th Edition PC/Windows XP.Copyright 2015. VSN International Ltd; 2015. https://www.vsni.co.uk/software/genstat[Accessed: 20/08/2018]

[19] Manavalan LP, Guttikonda SK, Nguyen VT, Shannon JG, Nguyen HT. Evaluation of diverse soybean germplasm for root growth and architecture. Plant and Soil. 2010;**330**:503-514

[20] Beebe SE, Rao IM, Blair MW, Acosta-Gallegos JA. Phenotyping common beans for adaptation to drought. Frontiers in Physiology. 2013;**4**:1-20. DOI: 10.3389/fphys.2013.00035

[21] Okogbenin E, Setter TL, Ferguson M, Mutegi R, Ceballos H, Olasanmi B, et al. Phenotypic approaches to drought in cassava: Review. Frontiers in Physiology. 2013;**4**:1-15

[22] Creelman RA, Mason HS, Bensen RJ, Boyer JS, Mullet JE. Water deficit and abscisic acid cause differential inhibition of shoot versus root growth in soybean seedlings. Plant Physiology. 1990;**92**:205-214

[23] Hsiao TC, Xu LK. Sensitivity of growth of roots versus leaves to water stress: Biophysical analysis and relation to water transport. Journal of Experimental Botany. 2000;**51**:1595-1616

[24] Bloom AJ, Chapin FS, Mooney HA. Resource limitation in plants—An economic analogy. Annual Review of Ecology and Systematics. 1985;**16**:363-392

[25] Sharp RE, LeNoble ME. ABA, ethylene and the control of shoot and root growth under water stress. Journal of Experimental Botany. 2002;**53**:33-37

[26] Tseng IC, Hong CY, Yu SM, Ho TTD. Abscisic acid- and stress-induced highly proline-rich glycoproteins regulate root growth in rice. Plant Physiology. 2013;**163**:118-134

Chapter 5

Tribological Behavior of Soybean Oil

Constantin Georgescu, Lorena Deleanu and George Catalin Cristea

Abstract

This chapter presents experimental data in the favor of using soybean oil, additivated or not, as lubricants, the market share of the soybean oil on the lubricants' market, a SWOT analysis for better configuring the tribological characteristics of the soybean oil and tribological parameters as friction coefficient, wear scar diameter, wear rate of wear scar diameter, etc. and their dependence on testing regime (load and speed). Also, the influence of temperature, shear rate, and oxidation parameters on the soybean oil viscosity is discussed.

Keywords: soybean oil, viscosity, tribology, friction, wear, scar wear diameter

1. About soybean oil, in the favor of using it as lubricant

1.1 SWOT and market analysis

The world vegetal oil production was around 182 million tons in 2016/2017 [1]. In 2017, the bio-lubricants market was evaluated at 2.47 billion USD, and it would value 3.36 billion USD by 2022, at a growth rate of 6.4%.

Environmental regulations, increasing production of vegetal oils, and their applications are market drivers. But their still high price and the decline in petrol resources are factors hindering the growth of this particular market [2]. North America is intended to be the largest market by 2022, but Europe led bio-lubricants market in 2016 and some of its member states (Germany, France, and Finland, but also Norway as an economic partner) ask for environment protection and biodegradability standards and regulations in the favor of these lubricants. For instance, the focus on "green chemistry", the regulation on environment protection and advances in research, should lead the bio lubricants to develop and gain the market on longer term, as an increase of using them is estimated if a more strong incentive policy is applied in European Union (**Table 1**).

The major restraints are deficiency in interconnecting regulations and higher prices than petroleum-based lubricants. The North American market for soy-based lubricants was estimated at USD 191.5 million in 2016 [4].

The Ag-Based Industrial Lubricants Research Center (USA, Northern Iowa) patented 30 genetically modified soybean lubricants (oils and greases) for tractor, chains, compressor, manufacturing processes and transmissions, metalworking and cooling fluids, fluids for the food industry, oils, transformer oils, greases for cars, railways, etc. [5].

IntechOpen

Industrial applications	2008	Production forecast 2020		
	Lubricants [t]	**Biolubricants [t]**	**Biolubricants, if a moderate incentive policy will act [t]**	**Biolubricants, if a strong incentive policy is applied [t]**
Hydraulic fluids	650,000	68,000	155,000	230,000
Oils for chainsaws	50,000	29,000	37,000	40,000
Mold release oils	100,000	9000	15,000	30,000
Other uses*	3,600,000	31,000	70,000	120,000
Total	**4,400,000**	**137,000**	**277,000**	**420,000**

**Lubricants for gears, engine, metal working, electric transformers, and base-stock for greases.*

Table 1.
Estimation of consumption of biolubricants [3].

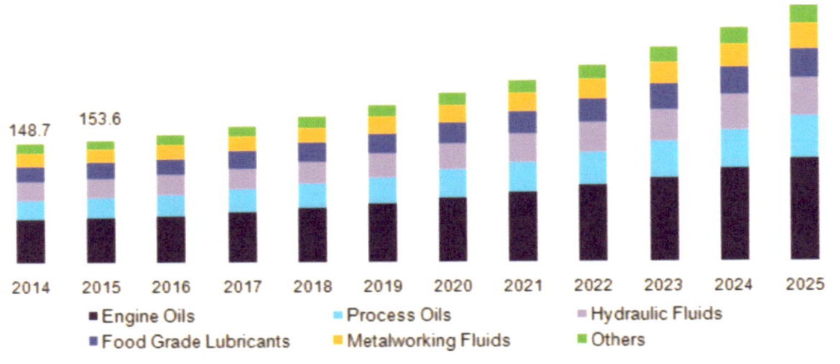

Figure 1.
Soybean lubricant market in the USA, depending on the application, for the time interval 2014–2025 (in million USD) [4].

A study of the American market reveals that the value of lubricants based on soybean oil will constantly increase, almost similar for major applications of these lubricants (engines, food processing, manufacturing process, and hydraulics) (**Figure 1**).

Research on soybean oil as gear lubricant was published by Ibrahim [6]. Soybean oil is used as biodiesel in a small share as compared to other vegetal oils, especially rapeseed oil, but the tribological characteristics are important even in this application [7].

Energy resources from renewable crops have gained prominence in order to replace petroleum products. Bio-lubricants become acceptable alternatives to conventional lubricants. Despite their benefits, these are still far from being practical. Since bio-lubricants are produced from raw vegetal oils, they have poor flow properties at low temperature and poor thermo-oxidative and hydrolytic stability. However, these shortcomings can be addressed by modifying the vegetal oils chemically [8] or incorporating additives into the oils [9, 10]. From **Figure 2**, one may notice that the production of soybean oil has been increasing constantly, and a considerable amount is used for producing lubricants.

Figure 3 presents the diversity in composition in fatty acids of several vegetal oils [12]. This would be the explanation of the very particular behavior of vegetal oils under boundary or fluid lubrication. Hence, it is necessary to control the fatty acid composition of vegetal oils [13]. The composition in fatty acids could also vary

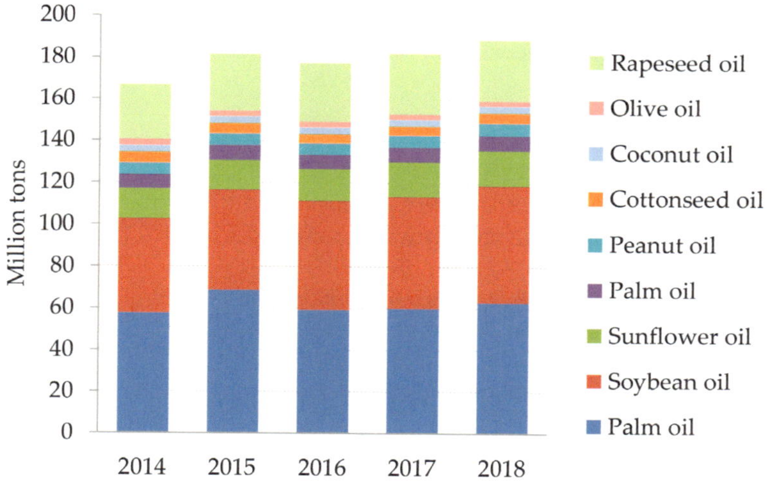

Figure 2.
Consumption of vegetal oils (adapted from [11]).

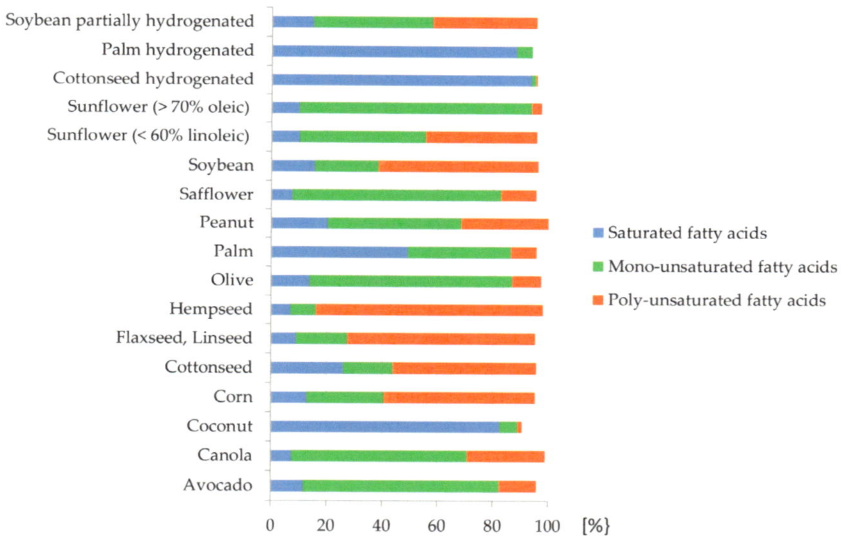

Figure 3.
Vegetal oils and their composition in fat acids [12].

with soil nature, climate, and human intervention on seeds, and even for the same place and the same type of seed, the annual conditions may influence the quality of vegetal oil.

Strengths, weaknesses, opportunities, and threats, well known as a SWOT analysis [14], for introducing vegetal oils in environmentally sensitive industries, agriculture, and transport, including soybean oil, as lubricants, underlines that a set of properties should be considered when the designer decides this lubrication solution [15–17].

Strengths in the favor of using soybean oil as oil for lubrication, either as neat, chemically modified or additivated, are discussed:

- biodegradability [18],
- a better lubricity, low volatility and a high viscosity index [17],

- better flammability characteristics (auto ignition points and higher ignition temperature on hot surfaces) than many mineral oils and similar to those of the rapeseed oil [19, 20],
- a good solubility for contaminants, additives, and polar deposits as compared to mineral oils [16, 17],
- environmental (nonpolluting or environmentally friendly) [21–23],
- extraction from renewable resources (even with a reference to 100 years) or the possibility of recycling or re-use of the lubricant [24].

Weak points for this vegetal oil include:

- lower viscosity as compared to mineral and synthetic oils [25–28],
- oxidation stability [9, 16, 19, 29–31],
- temperature range lower than that of mineral and synthetic oils [32],
- poorer properties at low temperature as compared to other lubricants [27, 29, 33].

The user should expect this vegetal oil to change its viscosity, oxidation stability, and polymerization during exploitation in a more intense way than mineral and synthetic lubricants. Chemical modification of soybean oil and/or the use of anti-oxidants [23, 34–36] could positively influence, but these will increase the cost of lubricant.

Opportunities are related to complying with more stringent environmental protection requirements that will minimize health and pollution risks. The new market shares for organic and biodegradable lubricants (obtained from renewable resources, especially plants) have increased for areas such as hydraulic fluids, chain lubricants, mold lubricants, two-stroke engines, turbine fluids, etc. [21].

Threats are the following:

- the need to redesign systems using bioliquids, a possibly costlier solution,
- accepting lowering some system operating parameters (especially load and maintenance, but not limited to) [22],
- the price still high (but not forgetting that, for example, synthetic oils in the 1990s were almost 10 times more expensive than mineral ones, today the ratio being only 3 to 1), market and users' inertia, the diversity of environmental and safety specifications, and a global policy that has not yet been clearly addressed on environmental issues.

1.2 Viscosity of soybean oil and soybean oil-based lubricants

For a lubricant to exhibit a better tribological behavior, it has to have an appropriate viscosity that will not decrease excessively when the working temperature increases. For many vegetal oils, their viscosity is low even at room temperature and it decreases dramatically when the oil is heated [19, 23, 26, 37].

Solea [19] did a comparative study for evaluating the viscosity of four vegetal oils, including soybean oil, experimental data proving the dependence of viscosity on temperature and shear rate (**Figure 4**), and tests done with the help of a rotational viscometer Rheotest2. The mineral oil OMV VG 46 was tested for comparison reason. In the temperature range 30–60°C, a more accentuated decrease of dynamic viscosity is noticed as compared to that characterizing the range 60–90°C. On the entire range of tested shear rates, the lowest decrease of viscosity was obtained for the corn oil (75.81%) and for the soybean oil (76.25%). For all tested vegetal oils, the dynamic viscosity decrease was 75–80%. This "agglomeration" of data may be the results of similar composition in fatty acids (see **Table 2**).

The dynamic viscosity of oils also decreases when temperature increases, for different shear rates (**Figure 5**), and for the lubricants tested in [19], the values have the tendency to agglomerate in the narrow range at higher temperatures (60–90°C) and for higher shear rates (**Figure 5**). This behavior has been also noticed in [38, 39].

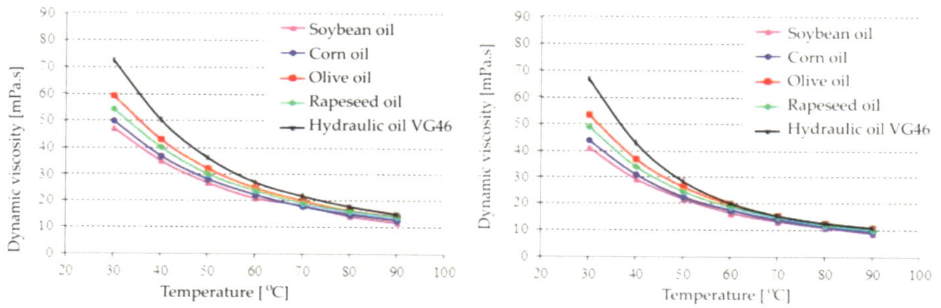

Figure 4.
Viscosity as a function of temperature for four vegetal oils and a hydraulic mineral oil for comparison reason (OMV ISO VG46), at two shear rate [19]. (a) Shear rate 10 s^{-1}. (b) Shear rate 80 s^{-1}.

Fatty acid	Symbol	Olive oil	Soybean oil	Corn oil	Rapeseed oil
Myristic acid	C14:0	—	0.11	0.05	0.05
Palmitic acid	C16:0	12.6	12.7	12.4	4.84
Palmitoleic acid	C16:1	1.20	0.13	—	0.06
Heptadecanoic acid	C17:0	0.10	0.05	0.12	0.14
Heptadecenoic acid	C17:1	0.10	0.06	0.05	—
Stearic acid	C18:0	—	5.40	2.10	0.14
Oleic acid	C18:1	79.30	21.60	28.45	62.73
Linoleic acid	C18:2	4.70	52.40	54.10	22.4
Linolenic acid	C18:3	0.80	5.70	1.10	7.50
Arachidic acid	C20:0	0.40	0.25	0.40	0.50
Eicosenoic acid	C20:1	0.25	0.20	0.35	1.25
Behenic acid	C22:0	—	0.50	0.10	0.30
Erucic acid	C22:1	—	0.16	—	—
Lignoceric acid	C24:0	0.16	0.20	0.10	—

Table 2.
Fatty acid composition of the tested oils [19].

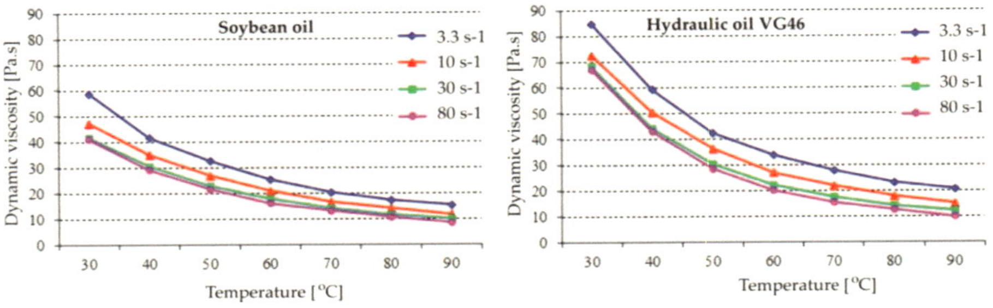

Figure 5.
Dynamic viscosity of soybean oil and a hydraulic mineral oil OMV ISO VG46, for different shear rates [19].

The temperature increase intensifies the intermolecular movement and reduces the attraction among oil molecules.

Solea [19] also measured the viscosity of soybean oil after oxidation, and data reveal a weak point of this vegetal oil: the oxidized soybean oil has an increasing viscosity with the time it bears oxidation (**Figures 6** and 7). The forced oxidation is realized by circulating air in the oil with a stable temperature.

Only a difference of 10°C of oil in the oxidation test (from 110 to 120°C) for tests during 10 hour modifies the dynamic viscosity of soybean oil, measured at 30°C

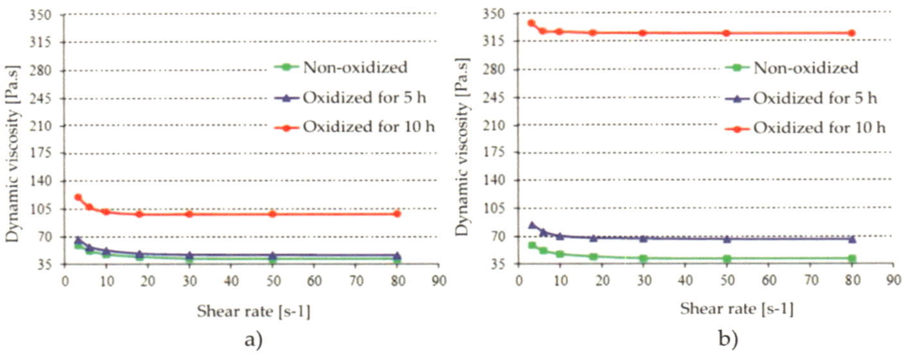

Figure 6.
Dynamic viscosity of soybean oil at 30°C, as a function of shear rate and oxidation time, after oxidation [14]. (a) After oxidation at temperature of 110°C. (b) After oxidation at temperature of 120°C.

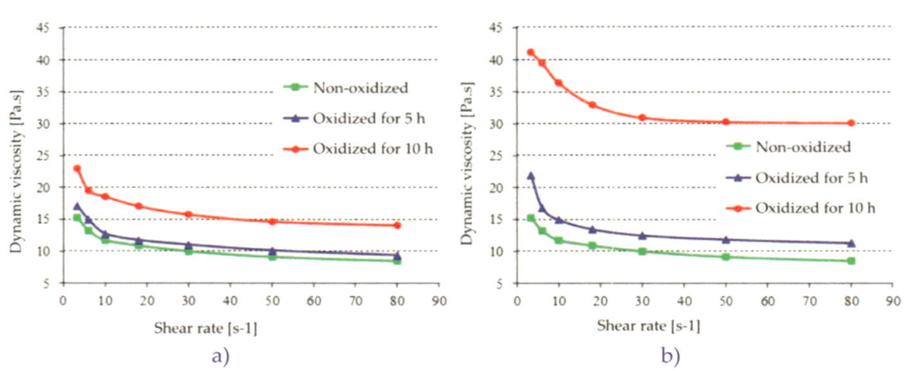

Figure 7.
Dynamic viscosity of soybean oil at 90°C as a function of shear rate and oxidation time (air flow 20 l/h in 25 ml of oil) [19]. (a) After oxidation at temperature of 110°C. (b) After oxidation at temperature of 120°C.

(**Figure 6**), but also at 90°C (**Figure 7**), with more than 300%, making the oxidized oil not to be recommended in applications where this oil has a working temperature more than 110°C and the oxidation could be generated (splash lubrication).

Figure 8 presents the influence of temperature on the dynamic viscosity of soybean oil when it is measured after oxidation during 5 and 10 hours, respectively, at constant temperatures of 110 and 120°C.

A similar tendency of evolution for soybean viscosity with shear rate and temperature obtained Esteban [40] and Cristea [41] but the latter for higher shear rates, using a Brookfield CAP 2000+ viscometer with cone 8 (**Figure 9**).

In terms of temperature viscosity dependence, the nanoadditives based on carbon (black carbon, graphite, and graphene) are separated in two groups:

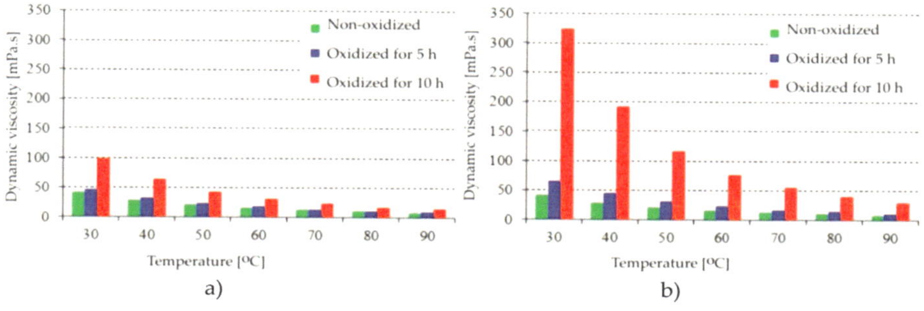

Figure 8.
Variation of dynamic viscosity of soybean oil, at shear rate of 80 s^{-1}, non-oxidized and oxidized at different temperatures: (a) Oxidation at constant temperature 110°C. (b) Oxidation at constant temperature 120°C [19].

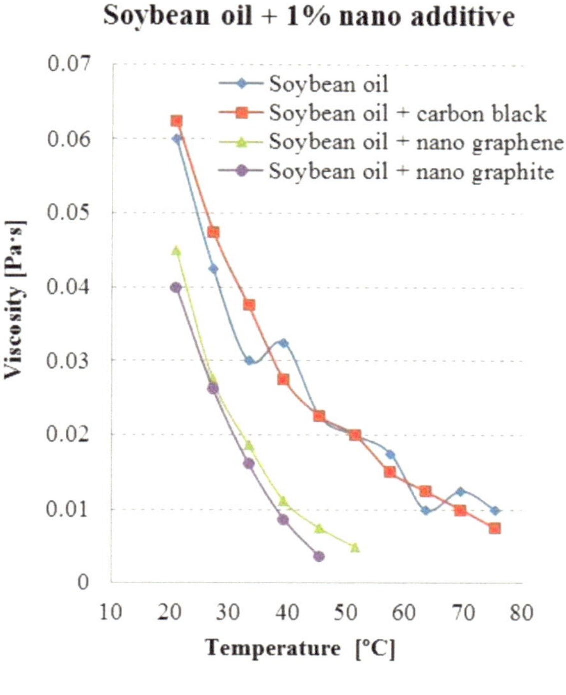

Figure 9.
Dynamic viscosity of soybean oil, nonadditivated, and additivated with 1% wt nanoadditive (tests done at shear rate 1000 s^{-1}) [41].

- black carbon does not significantly affect this dependence,
- nanographite and nanographene move down the curves of the dynamic viscosity dependence on temperature.

For any tested temperature and shear rate, the dynamic viscosity of soybean oil may be ordered [19]:

$$\eta_{\text{soybean oil}} < \eta_{\text{5 hour oxidation soybean oil}} < \eta_{\text{10 hour oxidation soybean oil}}$$

The increase in dynamic viscosity could be a criterion when evaluating the oxidation of vegetal oil. The soybean oil in modern applications especially needs a high degree of chemical stability of the lubricant.

2. Additivation of soybean oil

2.1 Classification of additives

Based on several relevant works in the literature [12, 42–45], the authors propose the classification given in **Table 3**.

Friction modifiers are adsorbed or fixed to the surface and form a film or a powdery intermediate layer that reduces friction. They can be classified into two distinct groups depending on the friction reduction mechanism:

- through the adsorbed film,
- by friction with the third body.

The first is generally due to polar molecules having a polar functional radical (alcohols, aldehydes, ketones, esters, and carboxylic acids) and a nonpolar terminal group. The polar group of the molecule adheres to the surface with long chains exposed to moving surfaces, reducing friction. They may also have polar elements that can chemically react with the surface to form a protective film. Vegetal oils and

Additives for lubricants			
Modifiers of chemical properties	**Modifiers of physical properties**	**Improvers of tribological behavior**	
		Antiwear and friction modifiers	**Extreme pressure additives**
Deposit control additives	Viscosity control additives	Inorganic	Phosphorus additives, like dialkyldithiophosphate (ZDDP), sulfur additives, sulfur–phosphorus additives, phosphorus–nitrogen additives, nitrogen additives, halogen additives, mixt additive package
Antioxidation additives	Poor point depressants	Organic	
Detergents	Antifoaming additives		
Toxicity	Dispersants		
Biodegradability			

Table 3.
A classification of additive for lubricants.

animal fats have such molecular structures, and therefore, they have good results in reducing friction.

Solid lubricants are added to vegetal oils with the same purpose of reducing friction and wear. This group includes not only carbon materials (fullerene, nanotubes, graphite, graphene, etc.) but also molybdenum and wolfram sulfides and fluorinated polymers, such as polytetrafluorethylene (PTFE) and perfluoropolyalkylethers (PFPAE). These can also be added in greases and composites that will function in dry conditions [46]. Solid lubricants (micro or nano) also help in situations where sliding surfaces have a rougher texture, "leveling" the profile of both surfaces. They are also recommended for reciprocal movements (in the case of the piston ring), which also produces a reduction in wear. They are added to lubricants that come into contact with surfaces with which EP (extreme pressure) additives cannot chemically react, such as polymers and ceramics and some of their composites [42].

Friction modifying and wear reducing additives can be grouped into solid lubricants and organic modifiers. The first group consists of carbon materials (graphite, graphene, black carbon, and fullerene), lamellar sulfides (tungsten and molybdenum), metal salts (boron nitride), and metal oxides (CuO, ZnO, and TiO_2 [47], which is not mentioned in [23]) but also linear polymers (polytetrafluoroethylene) [48]. Among the organic additives that act as friction modifiers are carboxylic acids or derivatives (stearic acid and esters), amides, imides, amines and their derivatives (oleyl amide, etc.), phosphoric and phosphonic acid derivatives, and organic polymers (methacrylates) [42, 43]. Regeneration of the friction reducing layer depends on additive concentration and conditions in which the tribosystem operates (speed, load, temperature, and contamination) [42].

Literature reported relatively low results on nanofluids as lubricants, mostly on transformer oil, silicon oil, gear oil, and heat transfer oil [49]. Limited investigations on the influence of nanoadditives on vegetal oils are presented [12]. Even if modern equipment working under high load, speed, and thermal conditions requires cooling and efficient lubrication, and for this concept, mineral and synthetic oils are still preferred and investigations on vegetal oils are needed for particular applications with environmental impact and in the perspective of oil resources extinction [50].

2.2 Specific processes for lubrication with nanoadditives

Wu et al. [43] propose a model that considers the lubricating additive concentration (see **Figure 10**). Although the model was created after experiments with TiO_2 as additive, it can be used to explain the behavior of lubricants with other nanoscale particles (metal 40 oxides, carbon materials, etc.). The fluid lubrication mechanism with nanoadditives has been also described in the works [52–54].

The mechanism for reducing friction and antiwear mechanism of nanoparticles in lubricants has been investigated, and it is based on the following processes [43]:

- micro-roll process [51, 55],
- process of forming a protective film [56–59],
- smoothing/leveling process [60],
- polishing process [51, 61], (**Figure 10**).

The first two mechanisms have a direct effect on lubrication [61]. In the case of rolling, no chemical reactions occur, and spherical or oval nanoparticles are willing

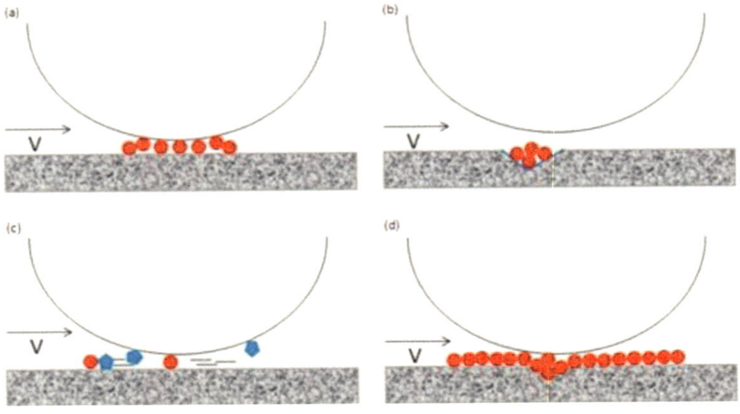

Figure 10.
The lubricating mechanism of water-based lubricants and TiO_2 as an additive. (a) rolling effect, (b) mending effect, (c) polishing effect, and (d) effect of protective film [51].

to roll. The lubrication mechanism of nanoparticles as friction modifiers includes three types of friction [62]:

- rolling friction—spherical nanoparticles act as micro or nano ball roller bearings between triboelement surfaces under light load conditions,
- sliding—nanoparticles serve as spacers and eliminate direct metal/metal contact between the asperities of the two triboelements, under higher load conditions,
- rubbing with the third body—exfoliating nanoparticles and their outer layers gradually transfer to surface texture, providing easier friction under high load conditions, when the third body can be considered a mixture of oil, nanoparticles, and wear particles.

The use of nanoparticles as lubricant additives is a top issue of research in the last decades [43, 63].

A spherical nanoadditive in contact [63] could act like a damper between two asperities in contact. It could change its shape becoming flatter when the load increases, thus, protecting a larger surface against rubbing (**Figures 11** and **12**).

Jayadas et al. [64] calculated the advantage of using additives in oils, based on the results of the shear rate and the temperature influence on the viscosity of the additivated lubricant. Wu et al. [51] reported increased load capacity of the additivated lubricants. Many studies were based on a single concentration of the additive. The effect of varying viscosity due to nanoparticle concentration is difficult to model, and therefore, tests become relevant.

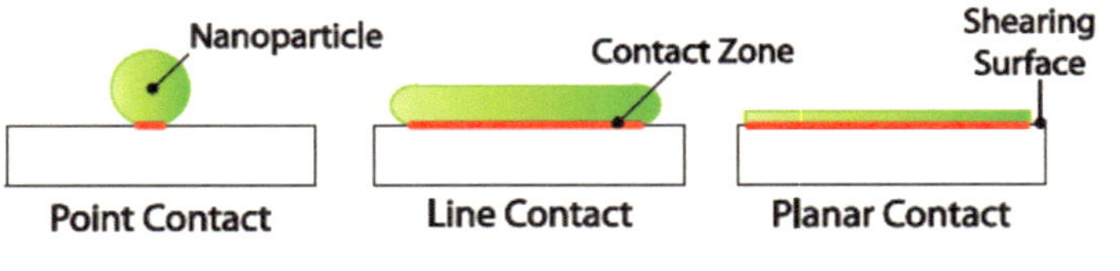

Figure 11.
Shape of a nanoparticle with the surface, under different operating regime [63]: punctual contact (low or no load), linear contact (mild load), and severe load (the move of surfaces one against another produces the shearing between additive and surface and not one among asperities).

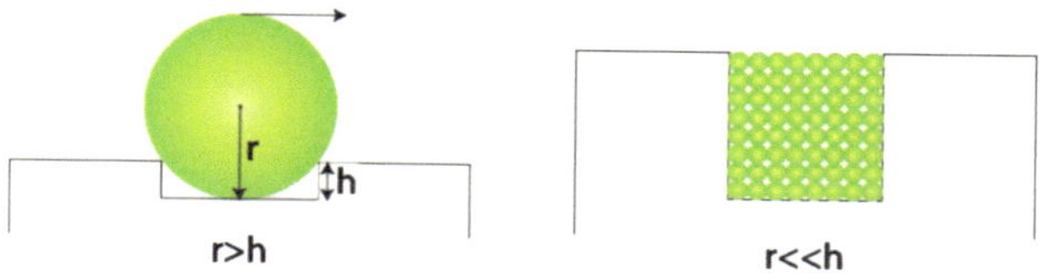

Figure 12.
The effect of relative size of surface texture and additive particles: particle with similar dimensions as the profile and particles smaller than the valleys of the texture [63].

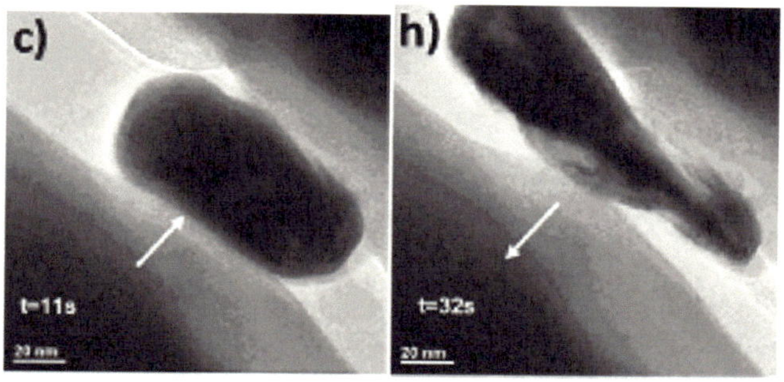

Figure 13.
Photos taken from the work of Lahouij et al. [65], the steps (here only two) through which a particle of WS_2 (wolfram disulfide) passes into a loaded contact.

A study by Lahouij et al. in 2012 [65] shows how the WS_2 (wolfram disulfide) particle protects the direct contact between metal asperities (**Figure 13**). The ovoid structure functions as a shock absorber and either the structure collapsed, or the particle was fragmented, it continued to remain between the two solid bodies. The hollow core of the particle was visible, and the deformation was large at the beginning of the stress, but as the load increased, the particle behaved like a variable-elasticity spring, the elastic characteristic actually increasing. Then, the particle begins to tear or scissor, and, finally, the WS_2 particle exfoliated in fragments.

From the studied literature, there is a tendency for deep research on additives in vegetal oils, especially those intended for lubrication. Attention should be focused on the dispersion of nanoadditive and the selection of dispersant.

3. Tribological characterization of lubricants based on soybean oil

3.1 Tribological parameters and testing equipment related to lubricants

In order to assess the quality of lubricants, it is important to establish the test methodology (equipment, parameters, and investigations during and after testing). Their selection depends primarily on the practical use for which the lubricant is tested, and the selected tribotester should approach as much as possible to the technical system in which the lubricant will be introduced.

Tribological tests can be grouped in severe tests and tests under normal working conditions. The here-presented results could be appreciated as moderate regimes.

Depending on the future application and the researchers' abilities and knowledge, tribological characterization will be done by a set of parameters, one being insufficient for this purpose. The evaluation of the results is often done by accepting a compromise, as durability and reliability of a system are influenced by synergic effects of actual dynamic parameters. There are presented several parameters that

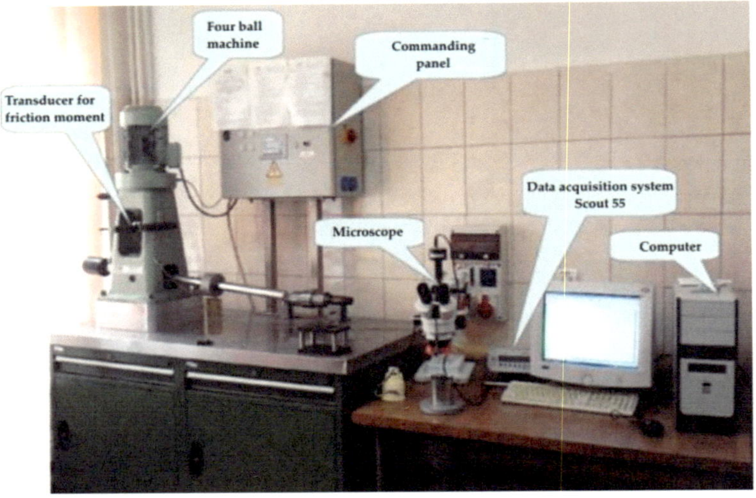

Figure 14.
Four ball machine ("Lubritest" Laboratory, "Dunarea de Jos" University of Galati).

could be taken into account for evaluating tribological behavior, with particular reference to four-ball tester (**Figure 14**), even if there are other tribotesters used for establishing the lubricating capabilities of vegetal oils: pin-on-disc, ball-on-disc, reciprocating rigs [66], etc.

The coefficient of friction (COF) may be analyzed by the following parameters:

- instantaneous value (i.e. at t time), paying attention to minimum and maximum values,
- mean value over the duration of the test (1 hour in this study, the number of samples per second being important in getting some peak values),
- average value over the last minutes of the test (argumentation: there are research reports presenting the average for 10 minutes, 5 minutes, as these time intervals are considered to be a stabilized domain),
- variation interval of the friction coefficient for 1 hour and for the last 10 minutes.

Several wear parameters for the tests done on four ball machine are given bellow:

Wear scar diameter (WSD) is the arithmetic average of the six diameter measurements, two on each of the three fixed balls of a test. For each ball, the wear diameter was measured in the direction of sliding and perpendicular to it. This value represents the diameter of the wear scar reported for each of the performed tests. The same method of obtaining the wear diameter is also given in specialized literature [66–68].

Wear as volume loss, considering the surface of wear scar as plane [69], is calculated as a sphere calotte having the diameter equal to the average wear scar diameter (see **Figure 15**):

$$V = \pi \cdot R^3 \left(1 - \sqrt{1 - \frac{1}{4}\left(\frac{D}{R}\right)^2}\right)^2 \left[1 - \frac{1}{3}\left(1 - \sqrt{1 - \frac{1}{4}\left(\frac{D}{R}\right)^2}\right)\right] \left[\mathrm{mm}^3\right] \qquad (1)$$

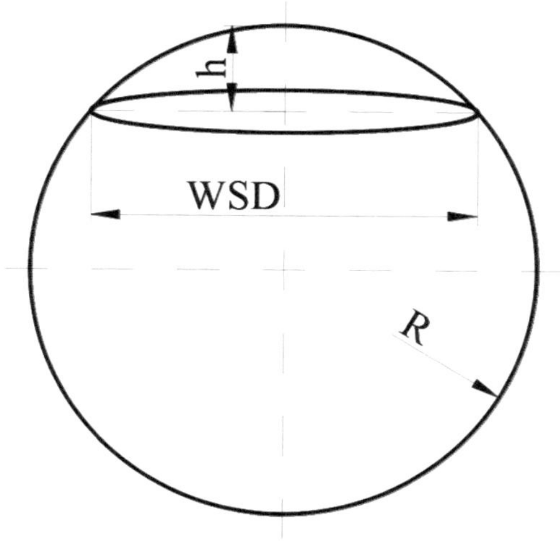

Figure 15.
The sphere calotte for calculating the worn volume on a ball.

where R is the ball radius; WSD is the wear scar diameter (calculated as average of six values of wear scars on fixed balls along sliding and perpendicular to it).

Wear rate of wear scar diameter is still rarely used, but it is more convenient for comparing results on four ball tribotester. Since the duration of the test is 1 h, the sliding distances are different for different speeds. For instance, for ball of 12.7 mm in diameter, tested 1 hour (this time being often selected by researchers), the sliding distance depends on the sliding speed in contact (that could be calculated knowing the rotational speed of the main shaft of the four ball machine): L(v = 0.38 m/s) = 1378.8 m; L(v = 0.53 m/s) = 1933.2 m; L(v = 0.69 m/s) = 2487 m. It is possible that the simple graph of the WSD dependence on additive concentration, load, and speed is not relevant due to the difference in the sliding distances, and then, on the basis of the literature [70], the wear can be also evaluated by another parameter called the wear rate, w:

$$w = \frac{\Delta V}{F \times L} \left[\mathrm{mm}^3/(\mathrm{N} \cdot \mathrm{m})\right] \tag{2}$$

where ΔV is the variation in sample volume (volume of removed material), F is the loading force; L is the sliding distance. The product F × L is the mechanical work done by the tribosystem; in other words, the wear rate shows the loss of material volume for the mechanical work unit performed by the system. The authors used a parameter named the wear rate of wear scar diameter, w(WSD), calculated as:

$$w(WSD) = \frac{WSD}{F \times L} \left[\mathrm{mm}/(\mathrm{N} \cdot \mathrm{m})\right] \tag{3}$$

where WSD is the wear scar diameter (calculated as average of six values of wear scars on fixed balls, along sliding and perpendicular to it), F is the load applied on the four balls, ***L*** is the sliding distance.

Since Blok had developed the concept of “flash-temperature” [71] in 1963, even critical comments on the constraints and limitations characterizing the model have accepted that this parameter is strongly depending on the local peak of the heat flow generated by friction [72]. Flash temperature parameter (FTP) is related to the critical flash temperature, above which a given lubricant does not operate in a convenient manner (it becomes ineffective) under the imposed conditions. Literature offers several relationships for calculating FTP, the authors selecting the one given by Marcher [72]:

$$FTP = \frac{F}{WSD^{1.4}} \left[\mathrm{N/mm^{1.4}}\right] \quad (4)$$

where F is the applied load on the four ball tribotester, in N; WSD is the average of the six measured wear scar diameters on the fixed balls in one test. Under constant load, there is an indirect proportionality between FTP and wear scar diameter.

FTP allows for evaluating the lubrication capacity of a fluid, especially under high loads, as these generate high temperature on rubbing surfaces, as it is the case of rolling mill and cutting processes. For a lubricant, FTP reflects the lowest temperature at which the liquid evaporates, risking auto-ignition in air. High values of this parameter are associated with a positive characteristic of a lubricant as it does not evaporate if the temperature in contact is low and the fluid film is thick enough to reduce friction and to avoid direct contact of asperities. Thus, the heat flow generated by friction will not be so high. These low temperatures and a reduced friction may also characterize boundary lubrication. Low values of FTP indicate the damage of the fluid film.

Oil film strength (OFS) is calculated using the load on a contact between the mobile and fix balls, Q , in N:

$$OFS = \frac{Q}{A_s} = \frac{0.408 \cdot F}{A_s} = \frac{1.632 \cdot F}{\pi \cdot WSD^2} \quad (5)$$

$$Q = \frac{F}{3\cos\theta} \cong 0.408 \cdot F \quad (6)$$

F is the applied load on the four ball tribotester, [N]; θ is the angle between the load direction on the main shaft of the four ball machine and the direction of normal load in the contact between the rotating ball and one fixed ball ($\theta \approx 35.264°$); A_s is wear area, calculated with the average WSD for the three fixed balls [mm^2].

Maps in tribological analysis are useful in assessing trends and determining test regimens for which two or more variables influence the tribological parameters; thus, the tribological behavior of the system is better revealed.

Investigations of the worn surfaces and used lubricants could be done with the help of FTIR (Fourier Transform Infrared) spectrometry, 3D profilometry [73, 74].

3.2 Tribological characteristics of soybean oil

Lubricant properties influence the tribological behavior of a system, one of the most important being the dynamic viscosity.

Georgescu presented a report [75] on different grade of rapeseed and soybean oils. The antiwear characteristics were tested for the following parameters: load on four ball tester—100 N, 200 N, and 300 N; sliding speed—0.46 m/s (1200 rpm), 0.57 m/s (1500 rpm), and 0.69 m/s (1800 rpm); test duration—60 minutes (rpm—rotations per minute of the main shaft of the four ball machine). Balls, as delivered by SKF (Swedish Ball Bearing Factory), are mirror-finished, with the arithmetic mean of absolute values of the ordinates z(x), measured from the mean line Ra = 0.02–0.03 μm and made of EN31 steel grade (also named 100Cr6) steel grade, having a hardness of 60–66 HRC and a diameter of 12.7 ± 0.0005 mm. Average values of friction coefficient (COF) for the 1 hour testing on four ball machine are acceptable for the tested soybean oils (below 0.1), but generally higher than those for mineral oil, a proof for a thinner film generated in contact at least for the tested parameters (**Figure 16**).

Comparing the wear rates of WSD for the tested oils, one may notice that the two soybean oils produced lower values than those obtained with the transmission mineral oil T90, for the load range 100–200 N (see **Figure 17**). At F = 100 N, both vegetal oils are acceptable for actual applications, but at F = 200 N, FTP is almost double for vegetal oils as compared to T90 and at high speed, the degummed soybean oil becomes competitive. At highest speed and load, values of FTP are closer, above 800 $N/mm^{1.4}$. T90 has a more pronounced increase of FTP with load and less with speed. This difference suggests the necessity of testing a lubricant and not to estimate by general considerations.

Based on four ball test data, Georgescu concluded that the vegetal oils could be acceptable and comparable with nonadditivated mineral oils like T90 [75]. The degumming process makes this soybean oil to generate a more intense wear, meaning that the eliminated substances would have contributed to a better tribological behavior. The problem is that the coarse oil is less stable in time and exposed to oxidation.

Values of COF are higher for the soybean oils suggesting a mixt or boundary regime, especially for load of 100 N. FTP was better, its values being greater than 1200 $N/mm^{1.4}$ for soybean oils, under the load of 200 N. For F = 100–200 N, FTP for T90 was in the range 550–680 $N/mm^{1.4}$. For F = 300 N, this parameter is kept at 1000–1200 $N/mm^{1.4}$ for all tested oils, only for speeds of 0.46 m/s and 0.57 m/s. The conclusion of the study presented by Georgescu [75] is that these two soybean oils could be used as lubricants for low loads and moderate speeds, and the degumming process does not influence significantly the tribological behavior, at least for the tested regimes, as evidenced by FTP values in **Figure 18** and by those for OFS in **Figure 19**. Each parameter has values in a narrow range for both coarse and degummed soybean oils. The higher difference in FTP for these soybean oils

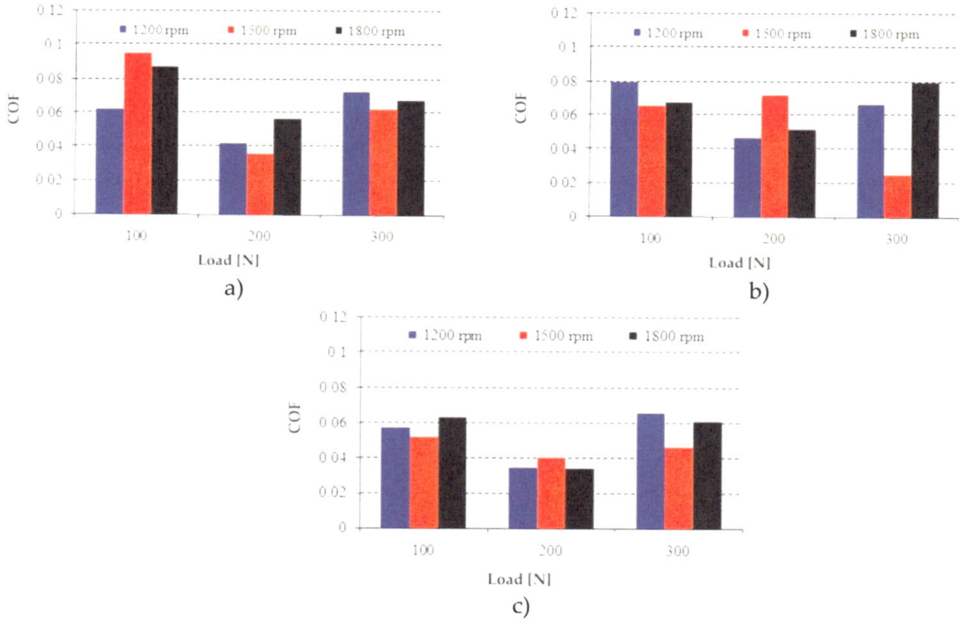

Figure 16.
Influence of load and sliding speed on the average value of COF [75]. (a) Coarse soybean oil (cold pressed). (b) Degummed soybean oil. (c) Transmission oil T90.

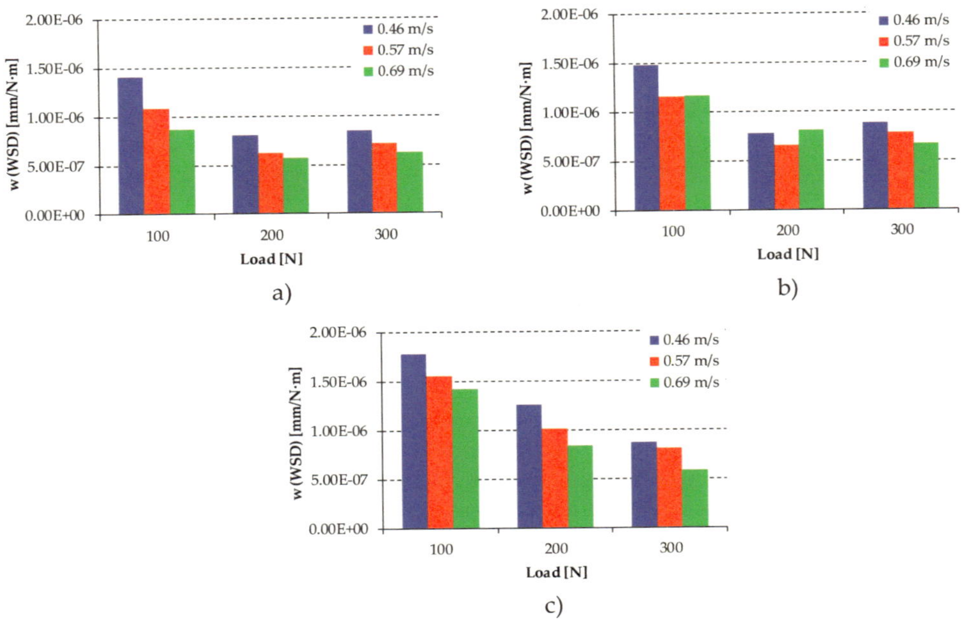

Figure 17.
Influence of load and sliding speed on the wear rate of WSD (wear scar diameter) [75]. (a) Coarse soybean oil (cold pressed). (b) Degummed soybean oil. (c) Transmission oil T90.

was noticed only for moderate load and that could be explained by the presence of gummy products in coarse soybean oil.

Cristea [41] reported the tribological behavior of soybean oil and soybean oil additivated with carbon-base nanoparticles, in different concentration (0.25, 0.5,

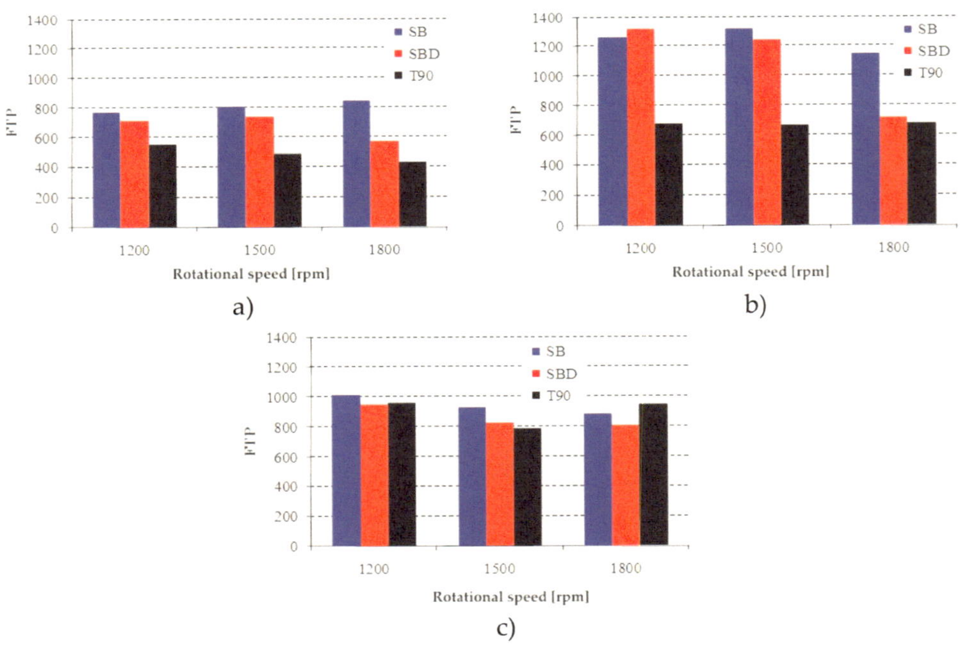

Figure 18.
FTP as function of load and sliding speed [75]. SB—coarse soybean oil, SBD—degummed soybean oil, T90—transmission oil. (a) 100 N. (b) 200 N. (c) 300 N.

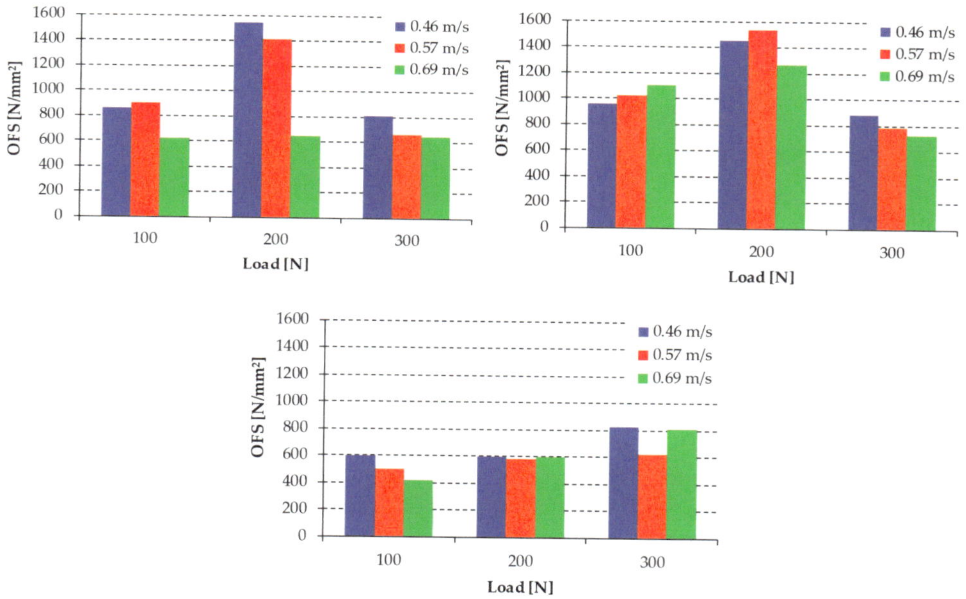

Figure 19.
Influence of load and sliding speed on the oil film strength (OFS) [75].

and 1%wt). **Figure 20** presents scanning electron microscopy (SEM images) for the nanoadditives that were supplied by PlasmaChem [76]:

- nanoamorphous carbon—average particle size ~13 nm, specific surface area ~550 m^2/g,
- nanographite—average particle radius 400–450 nm,
- graphene—nanoplates with a thickness of 1.4 nm and a particle size of up to 2 μm.

Figure 21 presents the evolution of COF over time, depending on load and speed, for two tests with the same parameters (F, v) when the four ball tribotester is lubricated with soybean oil. One may notice that, at the tested highest speed (v = 0.69 m/s), COF is less influenced by the applied load and performs in a narrow range meaning that speed is more important in generating a continuous fluid film, as argued by Dowson and Higginson for elasto-hydrodynamic lubrication [37].

a) b) c)

Figure 20.
Scanning electron microscopy for nanoparticles added in soybean oil [41]. (a) Black carbon. (b) Graphite. (c) Graphene.

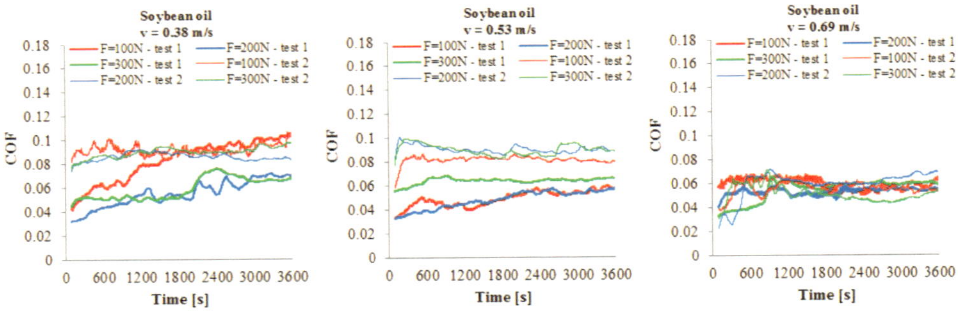

Figure 21.
The evolution of COF over time, depending on load and speed, for two tests with the same parameters (F, v) [41].

A representation of the wear rate of WSD (**Figure 22**) helps the researcher to observe the evolution trends of the parameter of interest according to two variables, here the tribotester load and the sliding speed of the rotating ball on the three fixed balls. For the range of analyzed loads and speeds, nonadditivated soybean oil had a downward trend with increasing load and speed. This trend is also consistent with Dowson and Higginson's argument on generating the elasto-hydrodynamic film [37] as wear will be reduced if the fluid film interposes between solid triboelements. They have shown that the speed factor $U = \eta_0(U_1 + U_2)/(2E' \cdot R_e)$ has the greatest influence on the minimum thickness of the fluid. For low-viscosity oils, the material factor $G\ (\alpha \cdot E')$ cannot participate in film formation to the same extent as the viscous oils at the working temperature of the contact [77]. (U_1, U_2 are the relative speeds of triboelements in contact, η_0 is the lubricat viscosity at areference temperature, E' is the equivalent Young modulus of the solid elements and, R_e is the equivalent radius of the contacting surfaces, α is the pressure-viscosity coefficient of the lubricant). In addition, vegetal oils are characterized by a high viscosity index, that is, the variation of this characteristic with the temperature is low, especially at temperatures above 50–60°C (see **Figure 4**). Images of wear scars in **Figure 23** point out a change of texture quality, especially when load increases to F = 300 N, even if the wear rate of WSD is the lowest for this load and all the tested speeds.

Cheenkachorn [78] studied three types of soybean oils (two given in **Table 4**, the third being an epoxidized soybean oil is simply the conventional soybean oil, in which all double bonds are epoxidized to form epoxide rings. Each of these oils was additivated with 1% ZDDP (Zinc dialkyldithiophosphates).

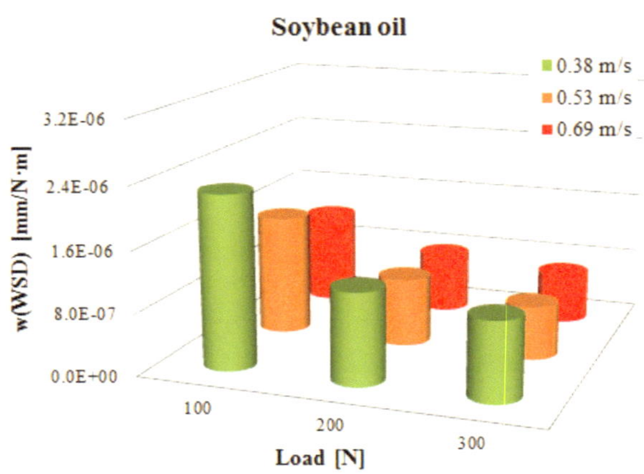

Figure 22.
Wear rate of WSD for the nonadditivated soybean oil [41].

The values of friction coefficient for all six oils (**Figure 24a**) show no clear trend. At 25°C and all speed conditions, epoxidized soybean oil without an antiwear additive has the highest friction coefficient. This is due to the fact that viscosity of

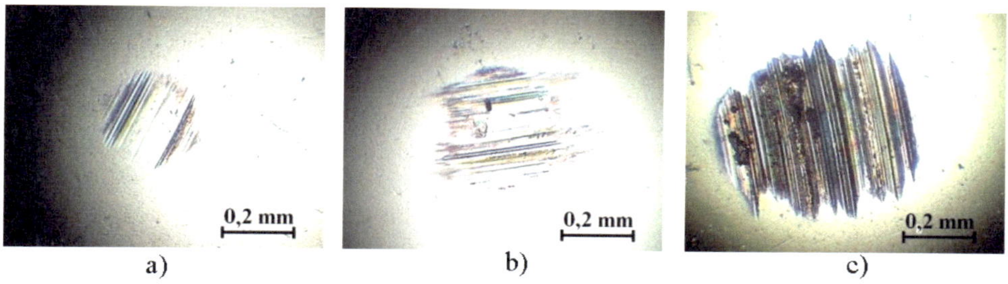

Figure 23.
The wear scars obtained with the soybean oil without additives v = 0.69 m/s [41].

Fatty acid	Symbol	Concentration, %wt			
		Refined soybean oil [41]	Soybean oil [78]	High oleic soybean oil [78]	Soybean oil [79]
Miritic acid	C14:0	0.11			0.1
Palmitic acid	C16:0	12.7	10.5	7	15
Palmitoleic acid	C16:1	0.13			0.15
Heptadecanoic acid	C17:0	0.05			0.15
Stearic acid	C18:0	5.40	4.1	4	6.78
Oleic acid	C18:1	21.60	23.4	83	26.6
Linoleic acid	C18:2	52.40	52.6	3	46.3
Linolenic acid	C18:3	5.70	7.2	2	2.69
Arachidic acid	C20:0	0.25			0.61
Gondoic acid	C20:1	0.20			
Eicosadenic acid	C20:2	0.50			

Table 4.
The characteristic fatty acid composition for the soybean oil, modified or not.

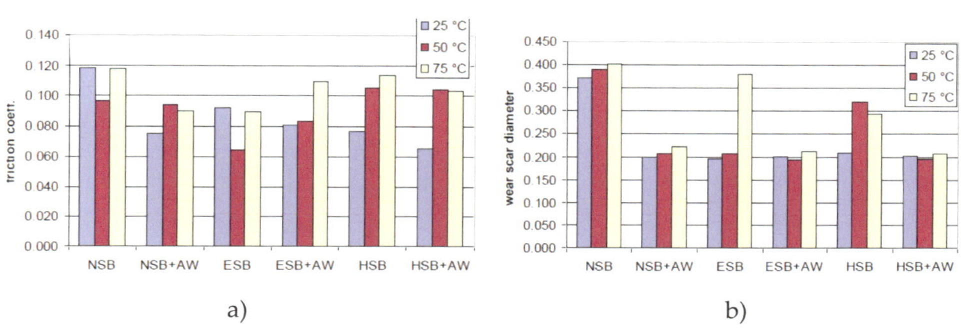

Figure 24.
Influence of different grades of soybean oil on friction coefficient (a) and wear scar diameter (b). Test conditions: 0.5 h, F = 110 N, v = 900 rpm, NSB—normal soybean oil, ESB—epoxidized soybean oil, HSB—high-oleic soybean oil, +AW—additivated with 1% ZDDP [78].

epoxidized soybean oil is higher than those of conventional soybean oil and high-oleic soybean oil. When the temperature increases, the viscosity of epoxidized soybean oil decreases. This results in a better oil circulation and forming of a multiple-layer film, which reduces the friction coefficient.

When comparing the results obtained by Cristea for refined soybean oil (see the first column in **Table 4**) with balls with very close characteristics [41], but for 1 hour test, 100 N at 1000 rpm, the higher value of WSD (0.413 mm) was just a little bit higher than that obtained by Cheenkachorn [78] for half testing time, meaning that during a longer time test, the wearing process is more intense especially in the beginning of the test, and the wearing process is slowing down due to lubricity of the lubricant (boundary lubrication) but also to the accommodation of surface textures in contact. ZDDP showed no clear influence on the trend of friction coefficient. For epoxidized soybean oil and high-oleic soybean oil, the temperature predominantly affected the wear scar diameters. This additive introduced in tested vegetal oils makes the temperature to have less influence on WSD (**Figure 24b**), and this being explained by the protection offered by the additive to the rubbing textures of the balls.

Zhao et al. [79] tested two types of oils with high viscosity, synthesized by nitrogen plasma polymerization of soybean oil. The nitrogen atoms were incorporated into the molecule of polymerized oil, and these three nitrogen heterocyclic compounds played a key role in improving tribological characteristics of polymerized oils. The lubricating properties of polymerized oils were tested on the four ball tester. The load-carrying capacities of polymerized oils reached 940.8 and 1049 N, respectively, higher than that of the unmodified soybean oil (646.8 N). They showed better antiwear properties under all tested loads and possessed preferable friction-reducing performances when the applied load surpassed 250 N. It was found that the nitrogen heterocyclic structure containing six atoms of nitrogen possessed higher coordination capacity than the ester groups of soybean oil and could form a durable organic nitrogen complex film on the metal surface. Simultaneously, the blended oils with different viscosity grades, which were prepared by diluting the polymerized oil with dioctyl sebacate, show excellent receptivity on the antiwear/extreme pressure additives of zinc dialkyl dithiophosphates and sulfurized isobutylene. Nitrogen plasma was used to open the C=C of soybean oil for polymerization. The values of kinematic viscosity (at 40°C) of the two polymerized soybean oils (PSO1 and PSO2) increased to 285 cSt (100 cSt = 1 cm^2/s) and 576 cSt from 33.8 cSt, respectively, and the viscosity indexes of PSO1 and PSO2 reached 220 and 283, respectively. The tribological characteristics (wear scar diameter and friction coefficient) are given in **Figure 25**, showing that the polymerization of soybean

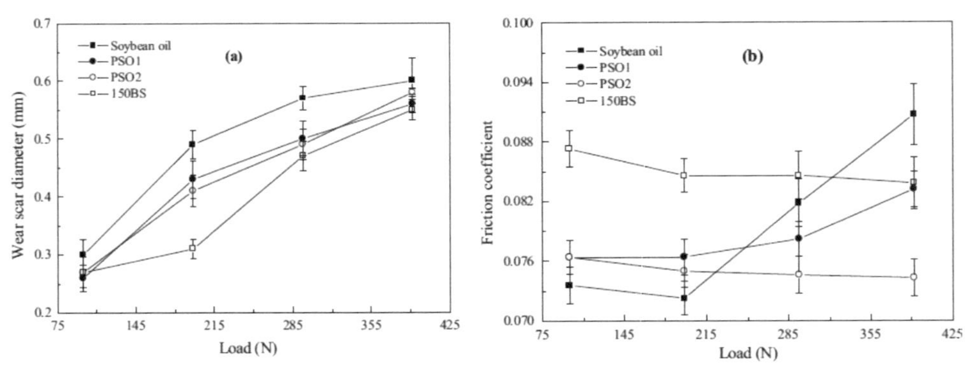

Figure 25.
Tribological characteristics of soybean oil, polymerized soybean oils (PSO1 and PSO2) and 150BS (industrial mineral oil with kinematic viscosity 601 cSt at 40°C and 31.4 cSt at 100°C, viscosity index 77) at different loads [79]: wear scar diameter (a), friction coefficient (b).

oil produces a decrease of WSD as compared to that produced by soybean oil. The friction coefficient seems to depend on the polymerization process, its values being less influenced by the load for the soybean oil PSO2. These better results for PSO2 may be explained by the nitrogen content, higher for PSO2, as result of a different time of polymerization.

3.3 Soybean additivation with carbonic materials

3.3.1 Laboratory formulations of additivated lubricants

The problem to be solved with such antiwear additives is their dispersion in oil. Cristea [41] proposed a method of obtaining a good dispersion taking into account that the tested base oil is a mixture of fatty acid triglycerides (see **Table 4**, first column). The steps followed in this laboratory technology were:

- mechanical mixing of the additive and an equal amount of guaiacol (supplied by Fluka Chemica) with the chemical formula $C_6H_4(OH)OCH_3$ (2-methoxyphenol) for 20 minutes; this dispersing agent is compatible with both the additive and the soybean oil (the mass ratio of the additive in the dispersing agent is 1:1, with an accuracy of 0.1 mg);

- gradually adding the soybean oil, measured to obtain 200 g of lubricant with the desired additive concentration (0.25%wt, 0.5%wt or 1%wt), by mixing with a magnetic homogenizer during 1 h;

- ultrasonication + cooling: 200 g of lubricant for 5 minutes using the Bandelin HD 3200 sonicator (Electronic GmbH & KG Berlin) sonicator; the lubricants are heated to about 70°C; the cooling time was 1 hour; this ultrasonic + cooling step is repeated five times to obtain a total of 60 minutes of sonication. The parameters of the ultrasonic regime are: 100 W power, frequency 20 kHz $\pm$ 500 Hz, continuous mode.

Figure 26.
Particles of nanoblack carbon on wear scar. Test conditions: v = 0.38 m/s, F = 200 N, time 1 h, lubricant: soybean oil +1% nanoblack carbon [47].

The method of sonication was also used for getting an acceptable dispersion of nanoparticles in lubricants by [80, 81].

3.3.2 Soybean oil additivated with nano carbon

Investigations by the help of scanning electron microscopy show that nanocarbon particles are on the friction surfaces as nanoagglomerations (**Figure 26**), on the surface texture of the wear scar. These particles or agglomerations appear to be rolled up and are likely to act as nanorolling elements, which explain low friction coefficients during the test (see **Figures** 27 and 28). The problem is that these particles are not uniformly distributed over the contact surfaces, producing a preferential wear on the particle-free areas. As the particles migrate in motion, these areas are prone to direct contact. This may be the explanation for the variation of the friction coefficient over time and with large amplitudes (**Figure 27**) and the variation of the average value of friction coefficient in larger ranges (**Figure 28**).

The friction coefficient plots of **Figure 27** are done using a moving average of 200 values, the recorded samples being of 2 values per second. The discussion of the

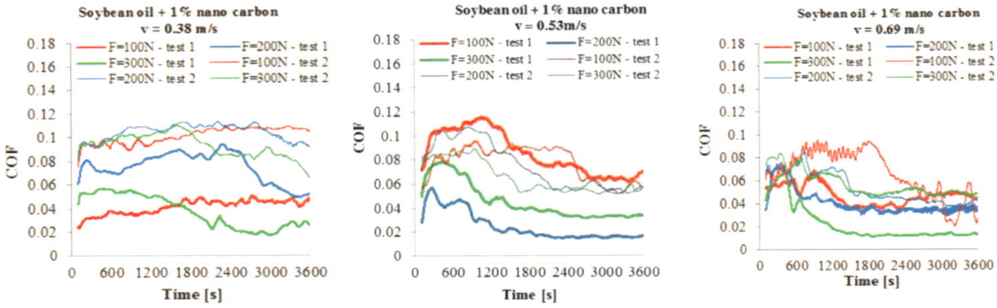

Figure 27.
The evolution of COF over time, depending on load and sliding speed for two tests with the same parameters (F, v) Cristea [82].

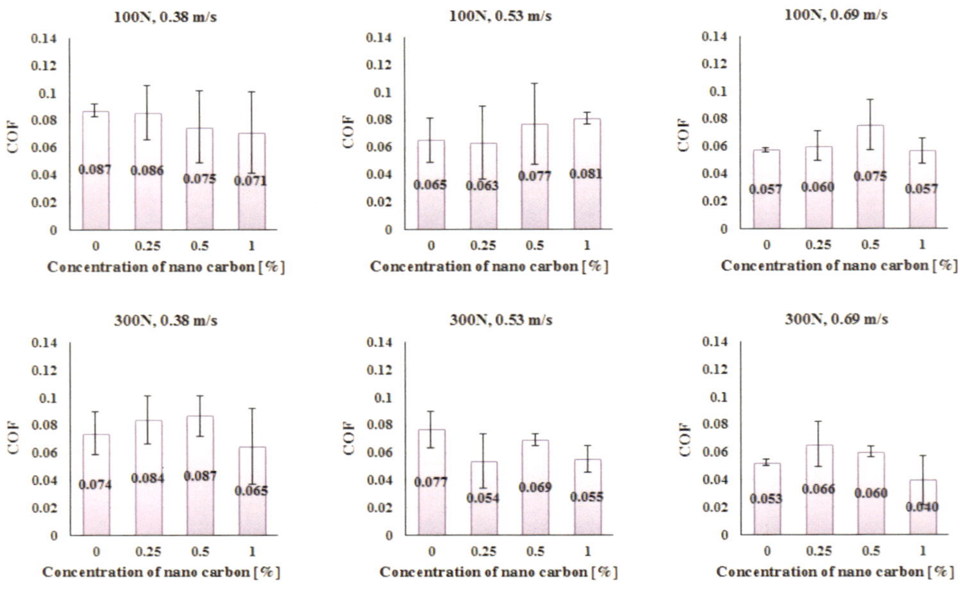

Figure 28.
Average values of friction coefficient (COF) during a test of 1 hour (two tests with the same parameters) [82].

evolution of the friction coefficient over time is based on the comments done by Czikos [66]. Thus, the coefficient for the soybean oil with nanocarbon is spread on a large interval for the lowest speed, but at v = 0.69 m/s, after a period with high values, COF performs in a narrow interval, under 0.06, meaning a full lubrication (**Figure 27**). For additivated lubricants, the tendency is to reduce the friction coefficient after a period of operation of 10–15 minutes.

Analyzing **Figure 28**, it is noted that at a concentration of 1.0% of nanocarbon, the friction coefficient becomes lower for higher load (F = 300 N) and high speed (v = 0.69 m/s). Under the minimum test load (F = 100 N), the COF oscillation range is the largest. Also, this regime gives less influence on wear rate of WSD (see **Figure 30**).

For nanocarbon additivated lubricants, average values of COF below 0.1 were obtained for all tests, except for the regime F = 100 N, v = 0.38 m/s, and v = 0.53 m/s and for F = 300 N, v = 0.38 m/s but just a little over 0.1. Addition of nanocarbon in soybean oil resulted in a COF decreasing trend for extreme test regimes ([F = 100 N, v = 0.38 m/s] and [F = 300 N, v = 0.69 m/s]). In the remaining combinations of test parameters, the influence of additivation on COF is not obvious.

This antiwear additive does not have a very clear influence on improving the tribological behavior of the soybean oil. Although the friction reduction mechanism exists in the presence of the additive, which is the interposition of carbon nanoparticles between the friction surfaces, as a third body friction, due to the migration of these particles (because they are not bonded to the surfaces) and to their uneven distribution in the contact, the tribosystem behaves more unstable than that using neat soybean oil. In a statistical approach, at some time moment and area of the contact, it could come in contact with particles sufficiently to reduce friction and wear, but during operation, there could be times when this number is low enough to have a mixt contact, and the oscillations between these two situations could explain the variations in friction coefficient and higher values for WSD.

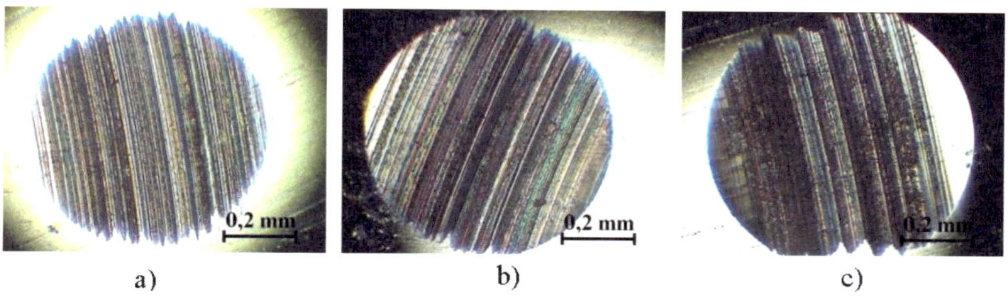

Figure 29.
Wear scar obtained with soybean oil +0.25% nanocarbon, F = 300 N (optical microscopy) [82]: v = 0.38 m/s (a), v = 0.53 m/s (b), v = 0.69 m/s (c)

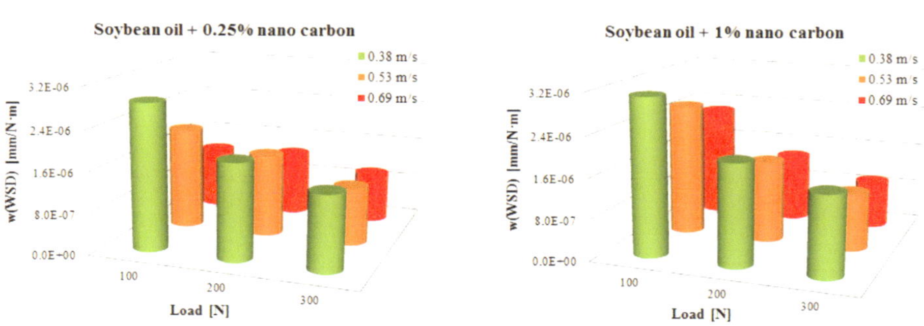

Figure 30.
Wear rate of wear scar diameter (WSD) for amorphous nanocarbon-additivated lubricants [41].

Since particle distribution is not even in contact during operation, this type of antiwear additive cannot help to improve tribological behavior because it does not reduce the friction coefficient and does not reduce the WSD as compared to those produce with neat soybean oil. The authors believe that the additive should be bonded (physically or chemically) for better results.

From **Figure 29**, it is noticed that the wear pattern did not increase too much with the speed, but the quality of the surface has considerably worsened, which justifies the profilometry study in [41, 73].

The wear rate of wear scar diameter, w(WSD), helps to determine the influence of the concentration of this nanoadditive. In graphs in Figure 30, the neat oil is not given. The additive, either with 0.25% or 1%, makes the wear parameter to visibly decrease with speed only for low load (F = 100N). Comparison with nonadditivated soybean oil is highlighted on the maps in **Figure 39**, where 0% additive concentration is for the neat oil. It can be noted a decrease of w(WSD) with load for all concentrations and speeds, for the additivated lubricants; the slope of the speed dependence for the same load is lower. At v = 0.38 m/s, the influence of the additive concentration is insignificant and the additivation would be justified in the field of high force for all speeds.

For nanocarbon additivated lubricants, w(WSD) is less sensitive to additive concentration, especially for F = 300 N. The nonadditivated soybean oil can be recommended for light regimes (equivalent to F = 100–200 N and speed v = 0.38–0.69 m/s). The almost linear dependence of WSD on the concentration of this additive is only observed for combinations with F = 100 N. For the tested regimes (F = 100–300 N and v = 0.38–0.69 m/s), the results are not in the favor of nanocarbon additivation of the soybean oil.

3.3.3 Soybean oil additivated with nanographite

Figure 31 shows the evolution over time of COF for all tests performed with soybean oil additivated with nanographite. There is a narrowing of its evolution range for v = 0.69 m/s for all loads and a scattering of higher COF values for low speeds and loads.

Analyzing **Figure 32**, it can be noticed that, at F = 100 N (first horizontal line), the nanoadditive does not dramatically alter COF average. At high load (F = 300 N), COF increased for all additivated soybean oils as compared to the neat oil. The explanation would be that the graphite does not cover the entire surface of the contact but is only present in contact in the form of nanorolls, the reduced friction being zonal. There are also direct friction areas and friction areas with the third body (where nanoparticles or microparticles generated by agglomerating the first

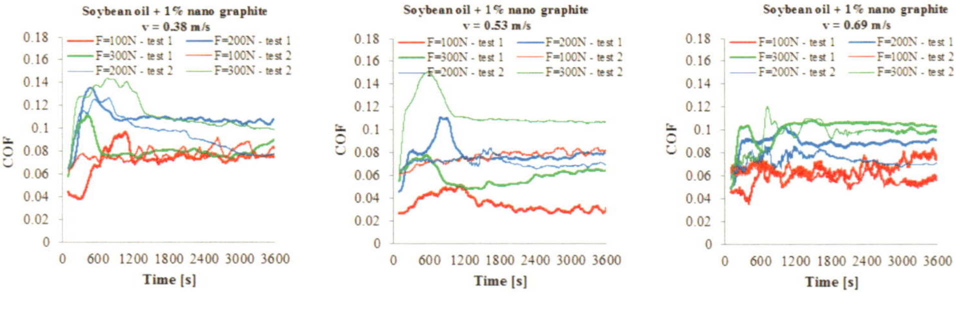

Figure 31.
The evolution of COF over time, for different loads and speeds, for two tests with the same parameters (F, v) [82].

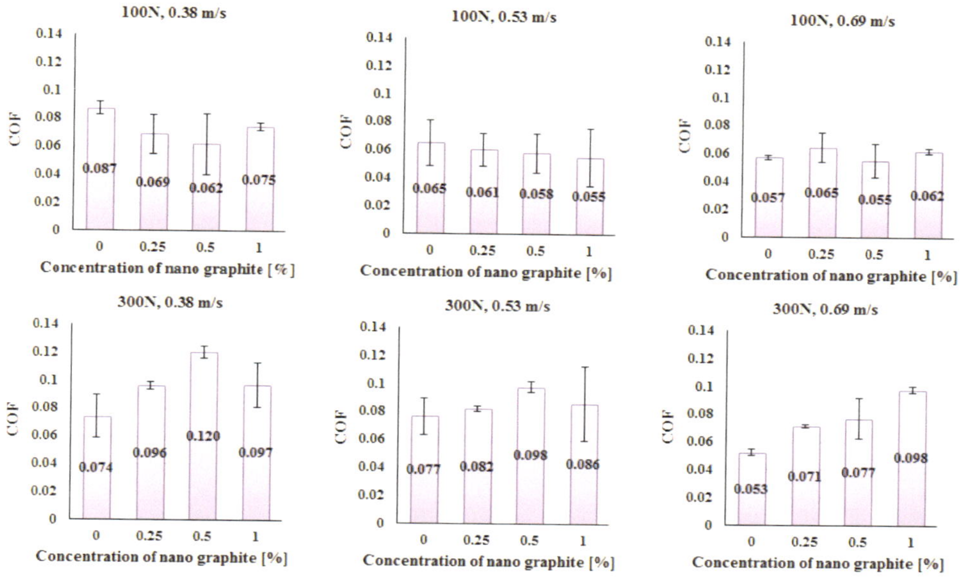

Figure 32.
Average values and scattering intervals for two tests performed with the same parameters (F, v, c) [82, 83].

ones due to load and surface texture). It appears that the presence of graphite prevents the generation of EHL (elasto-hydrodinamic lubrication) film as COF has higher values, toward 0.1, especially for F = 300 N. No lower average COF values than those for the soybean oil have been obtained, except for tests: (F = 100 N, v = 0.38 m/s) and (F = 100 N, v = 0.53 m/s), with a graphite concentration of 0.25%wt. But the differences are too small to highlight an influence of the additive or the test regime. High values for wear rate of WSD at low load and speed imply more intense abrasion, which occurs if the lubricant film does not form and/or if the additive does not protect the contact. Maybe local particle agglomerations make the friction coefficient to oscillate and, when they migrate in contact, they allow one triboelement to fall over the other, in direct contact under higher load than if it had not encountered the graphite agglomerations.

Analyzing the photos in **Figure 33**, it can be noticed that the nature of the wear pattern does not change significantly, resulting from the abrasive wear process and with rare adhesive wear spots at higher loads.

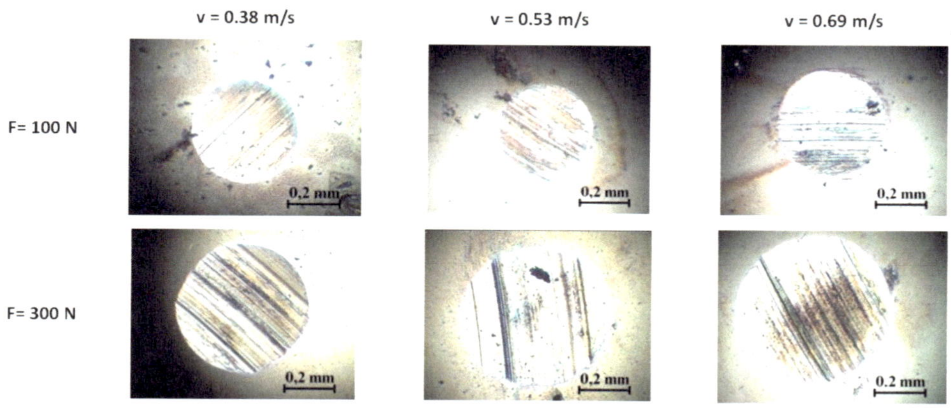

Figure 33.
Photos of the wear scars of the soybean oil +1% nanographite [82].

Comparing the graphs in **Figure 34**, they are similar in appearance, regardless of the concentration of the nanoadditive. The wear rate decreases with increasing load; for load F = 300 N, the wear rate of WSD is less influenced by speed.

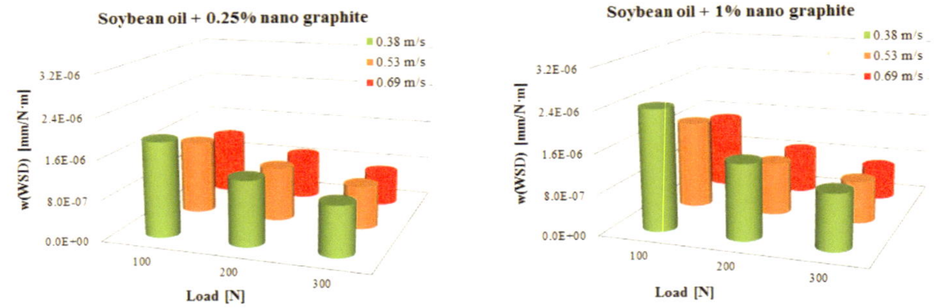

Figure 34.
Wear rate of WSD for nanographite additivated lubricants [82].

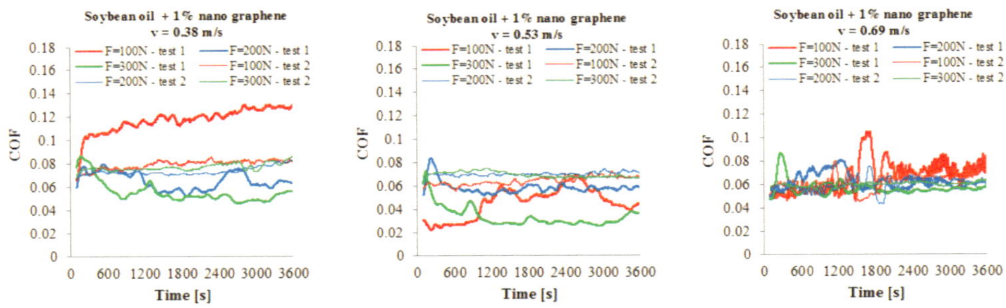

Figure 35.
The evolution of COF in time, depending on load and speed, for two tests with the same parameters (F, v) [41].

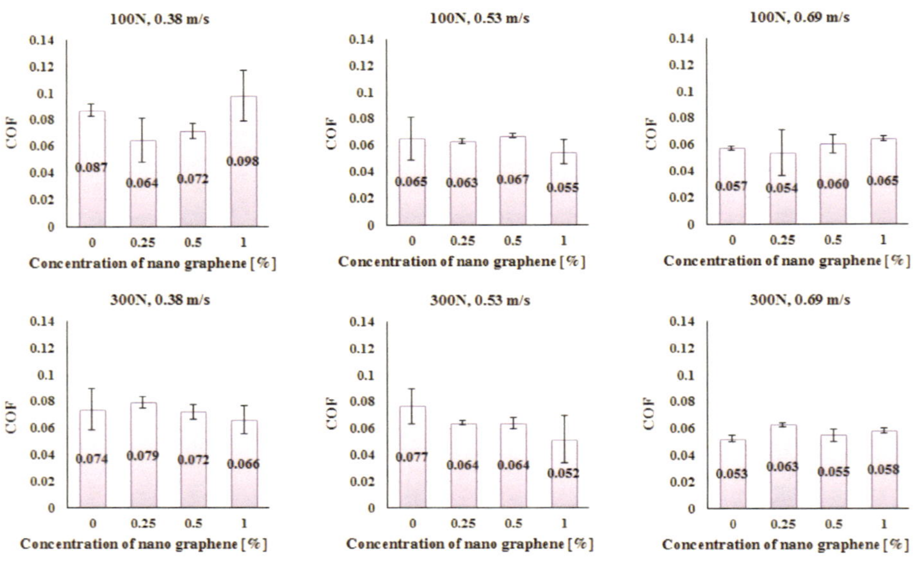

Figure 36.
Average values and spread range of friction coefficient (COF) for two tests performed with the same parameters (F, v, C) [41].

3.3.4 Soybean oil additivated with graphene

The evolution of COF over time is given in **Figure 35**, better ones being obtained for the highest concentration. COF variations have shortcuts or growth levels, which can be explained by the dynamics of COF components (dry friction, third body rubbing in areas with graphene nanoparticles, and partial fluid friction).

The addition of graphene does not improve COF but keeps it very close to the values of neat soybean oil. At v = 0.38 m/s, the highest values were obtained irrespective of the concentration of the additive, suggesting that the improvement in friction (in the sense of reducing it) is due to the increase in speed (**Figure 36**) [41] and not on the additive, but the graphene does not prevent the formation of the fluid film.

WSD does not significantly increase but the texture of the surface visibly changes (**Figure 37**), and the wear rate of WSD indicates a better tribological behavior of additivated soybean oil with graphene, but for more severe regimes (F = 200–300 N and v = 0.53–0.69 m/s), from this set of observations, the importance of correlation in the interpretation of several tribological parameters is pointed out.

High values were obtained for the mildest test regime (F = 100 N, v = 0.38 m/s). One could argue that a low loaded contact does not keep the additive in contact (pressed and hung on the texture). The lowest value for the most severe regime (F = 300 N, v = 0.69 m/s) was explained by forming the EHD film and maintaining the nanoadditive in contact.

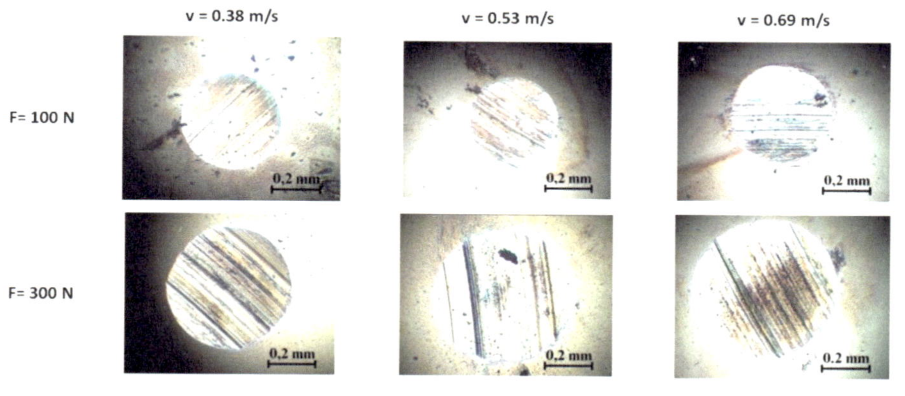

Figure 37.
Optical microscope photos of wear scar diameter, after testing with soybean oil +0.5% nanographene [41].

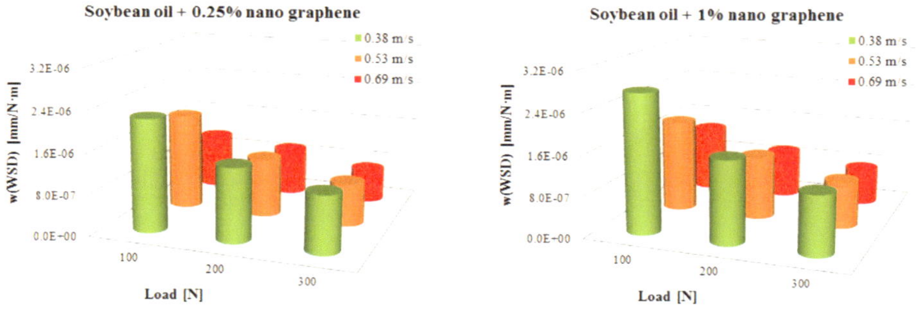

Figure 38.
w(WSD) of soybean oil additivated with nanographene [41].

The wear rate of WSD (**Figure 38**) is similar for 0.25 and 1%, meaning the additive concentration (0.25–1%) does not influence to much the wear. It seems that the wear is smaller for longer sliding distances and higher sliding speeds.

3.4 Maps for tribological parameters

The influence of the quality of the additive is manifested not only by the minimum value but also by the map area for which the minimum values of w(WSD) are spread (**Figure 39**). The lower surface of the map was noticed for graphene at 1% wt, between F = 200 N and F = 300 N and v = 0.69 m/s, less influenced by the amount of additive. For carbon, the low wear rate area is narrower. At v = 0.38 m/s and the lowest tested load, the lowest value of w(WSD) is obtained for graphite. The concentration of 1% nanoadditive enlarges the domain of reduced wear rate, meaning the load and speed have less influence on the test regime, especially for higher values for both parameters.

Maps were represented using a spline interpolation, and the surfaces are "compelled" to include the experimental data. A point on a wear rate map area is the wear rate of WSD for a test characterized by the set of input parameters (F [N], v [m/s],

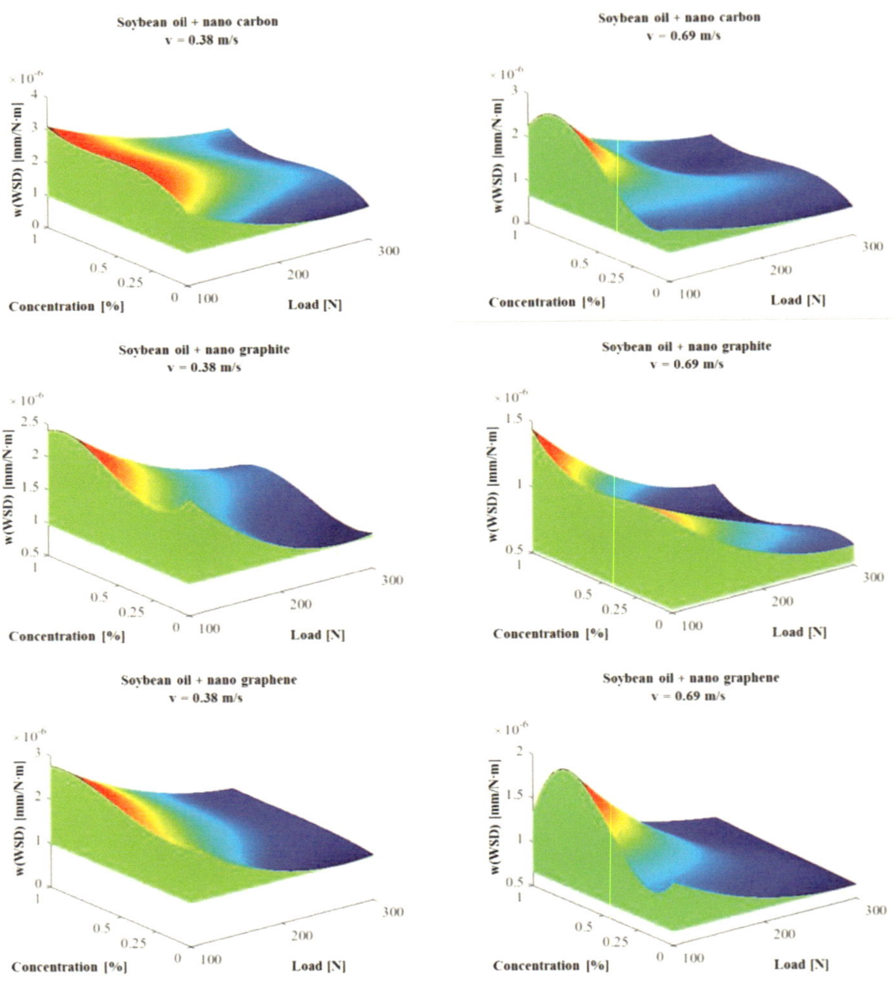

Figure 39.
Maps of w(WSD) for lubricants tested at two sliding speeds [41].

and c [%]), where F is the normal load on the four ball tribotester, v is the sliding speed, and c is the mass concentration of nanoadditive.

4. Conclusions and new trends in using soybean oil as lubricant

Friction coefficient and wear rate of wear scar diameter are important tribological characteristics, and they are compared in **Figures 40** and **41** for neat soybean oil and the same oil additivated with nanocarbonic additives. COF has higher average values for soybean oil with graphite. This fact could be explained by local agglomerations of the nanosheets of graphite that migrate in contact, causing a mixt regime especially when the COF values overpass 0.1. Agglomeration of nanoparticles could also explain the balls. The high values in light regimes could be explained by the fact that the particles are not pressed enough to fill and remain on the surface texture).

The downward trend of wear rate of wear scar diameter, w(WSD), had a higher gradient for lubricants with 1% nanoadditive (**Figure 41**), which would recommend further testing for more severe regimes, where additives are likely to better protect the surface of the contact.

Analyzing these tribological parameters, the authors consider that a combination (a low and constant evolution in time of friction coefficient, a small WSD, and a high value of FTP) makes the lubricant to have a good reliability in functioning. These laboratory test results have to be carefully applied when designing an actual applications with such a lubricants as in practice, the range of parameter variations is larger because of perturbations like vibrations, mechanical shocks, operator's errors, humidity, higher temperature gradients, etc.

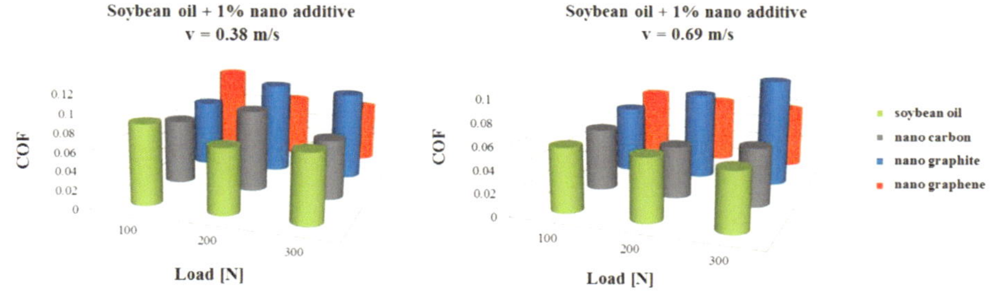

Figure 40.
Influence of additive in concentration of 1% on the friction coefficient [41].

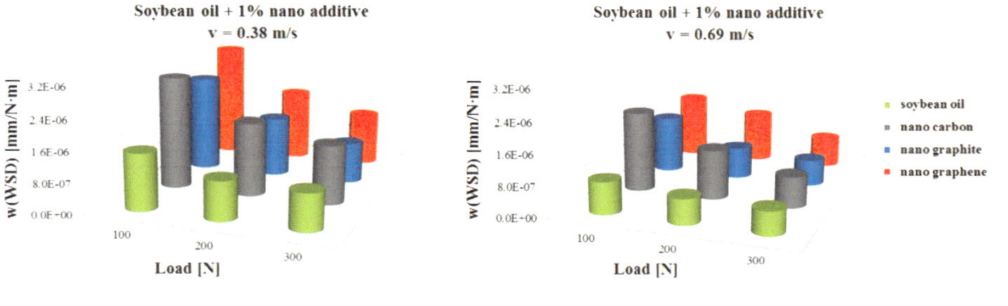

Figure 41.
Maps of the wear rate of WSD for lubricants additivated with 1% nanoadditive [41].

Author details

Constantin Georgescu[1], Lorena Deleanu[1*] and George Catalin Cristea[2]

1 Department of Mechanical Engineering, Faculty of Engineering, “Dunarea de Jos”, University of Galati, Romania

2 National Institute for Aerospace Research "Elie Carafoli"—INCAS, Bucharest, Romania

*Address all correspondence to: lorena.deleanu@ugal.ro

References

[1] Nutrient Database, Release 24 [Internet]. 2011. Available from: https://www.ars.usda.gov/ARSUserFiles/80400525/Data/SR24/sr24_doc.pdf [Accessed: 2018-02-21]

[2] Bio-Lubricant Market by Base Oil Type (Vegetable Oil, Animal Fat), Application (Hydraulic Fluids, Metalworking Fluids, Chainsaw Oils, Mold Release Agents), End Use (Industrial, Commercial Transport, Consumer Automobile)—Global Forecast to 2022 [Internet]. 2018. Available from: https://www.researchandmarkets.com/research/njfr5x/biolubricant?w=5 [Accessed: 2018-04-10]

[3] Biolubricants: The sector gradually organises itself [Internet]. 2012. Available from: http://www.agrobiobase.com/en/article/biolubricants-sector-gradually-organises-itself [Accessed: 2018-04-10]

[4] North America soybean oil-based lubricants market analysis, by application (metalworking fluids, engine oils, hydraulic fluids, process oils), by country, and segment forecasts, 2014–2025 [Internet]. 2017. Available from: http://www.grandviewresearch.com/industry-analysis/north-america-soybean-oil-based-lubricants-market [Accessed: 2018-04-10]

[5] The National Ag-Based Lubricants Center & Transportation Sustainability [Internet]. 2006. Available from: http://www.intrans.iastate.edu/mtc/documents/tsPresentations/2006seminar/james_mar31.pdf [Accessed: 2018-04-11]

[6] Ibrahim A, Ishak SSM, Kamaruddin MF. Comparison between sunflower oil and soybean oil as gear lubricant. Applied Mechanics and Materials. 2015;**699**:443-448. DOI: 10.4028/www.scientific.net/AMM.699.443

[7] Peng D-X. Room temperature tribological performance of biodiesel (soybean oil). Industrial Lubrication and Tribology. 2016;**68**(6):617-623. DOI: 10.1108/ILT-10-2015-0143

[8] Adhvaryu A, Erhan SZ. Epoxidized soybean oil as a potential source of high-temperature lubricants. Industrial Crops and Products. 2002;**15**:247-254. DOI: 10.1016/S0926-6690(01)00120-0

[9] Castro W, Perez JM, Erhan SZ, Caputo F. A Study of the oxidation and wear properties of vegetable oils: soybean oil without additives. Journal of the American Oil Chemists' Society. 2006;**83**:47-52. DOI: 10.1007/s11746-006-1174-2

[10] Fang JH, Xia DY, Chen BS, Wu J, Wang J. Friction and wear performances of magnesium alloy against steel under lubrication of soybean oil with s-containing additive. Applied Mechanics and Materials. 2014;**538**:19-23. DOI: 10.4028/www.scientific.net/AMM.538.19

[11] Consumption of vegetable oils worldwide from 2013/14 to 2017/2018, by oil type [Internet]. 2017. Available from: https://www.statista.com/statistics/263937/vegetable-oils-global-consumption/ [Accessed: 2018-04-11]

[12] Darminesh SP, Sidik NAC, Najafic G, Mamat R, Ken TL, Asako Y. Recent development on biodegradable nanolubricant: A review. International Communications in Heat and Mass Transfer. 2017;**86**:159-165. DOI: 10.1016/j.icheatmasstransfer.2017.05.022

[13] Biresaw G, Bantchev G. Effect of chemical structure on film-forming properties of seed oils. Journal of Synthetic Lubrication. 2008;**25**:159-183. DOI: 10.1002/jsl.58

[14] Helms MM, Nixon J. Exploring SWOT analysis—where are we now?: A review of academic research from the last decade. Journal of Strategy and Management. 2010;**3**:215-251. DOI: 10.1108/17554251011064837

[15] Cermak SC, Biresaw G, Isbell TA, Evangelista RL, Vaughn SF, Murray R. New crop oils: Properties as potential lubricants. Industrial Crops and Products. 2013;**44**:232-239. DOI: 10.1016/j.indcrop.2012.10.035

[16] Erhan SZ. Industrial Uses of Vegetable Oils. Peoria: AOCS Press; 2005. 184 p. ISBN: 1-893997-84-7

[17] Rudnick LR, Erhan SZ. Natural oils as lubricants. In: Rudnick LR, editor. Synthetics, Mineral Oils and Bio-Based Lubricants. Chemistry and Technology. Boca Raton: CRC Press; 2006. pp. 353-360. ISBN: 1-57444-723-8

[18] Luna FMT, Rocha BS, Rola JEM, Albuquerque MCG, Azevedo DCS, Cavalcante JCL. Assessment of biodegradability and oxidation stability of mineral, vegetable and synthetic oil samples. Industrial Crops and Products. 2011;**33**:579-583. DOI: 10.1016/j.indcrop.2010.12.012

[19] Solea LC. Contributii la studiul comportarii reologice si tribologice a unor lubrifianti biodegradabili pe baza de uleiuri vegetale [thesis]. Galati: Dunarea de Jos University; 2013

[20] Georgescu C, Cristea GC, Solea LC, Deleanu L, Sandu IG. Flammability of some vegetal oils on hot surface. Revista de Chimie. 2018;**69**(3):668-673

[21] Norrby T, Kopp M. Environmentally adapted lubricants in Swedish forest industry: A critical review and case study. Industrial Lubrication and Tribology. 2000;**52**(3): 116-125. DOI: 10.1108/00368790010326438

[22] Nagendramma P, Kaul S. Development of ecofriendly/biodegradable lubricants: An overview. Renewable and Sustainable Energy Reviews. 2012;**16**:764-774. DOI: 10.1016/j.rser.2011.09.002

[23] Miller M. Additives for bioderived and biodegradable lubricants. In: Rudnick LR, editor. Lubricant Additive. Chemistry and Applications. 2nd ed. Boca Raton: CRC Press; 2009. pp. 445-454. ISBN: 978-1-4200-5964-9

[24] Bio-based industries, towards a public-private partnership under Horizon 2020? [Internet]. 2013. Available from: http://www.gppq.fct.pt/h2020/_docs/brochuras/bioeco/bio-based-industries_en.pdf [Accessed: 2018-04-11]

[25] Paredes X, Comunas MJP, Pensado AS, Bazile J-P, Boned C, Fernández J. High pressure viscosity characterization of four vegetable and mineral hydraulic oils. Industrial Crops and Products. 2014;**54**:281-290. DOI: 10.1016/j.indcrop.2014.01.030

[26] Biresaw G, Bantchev GB. Pressure viscosity coefficient of vegetable oils. Tribology Letters. 2013;**49**:501-512. DOI: 10.1007/s11249-012-0091-9

[27] Liu Z, Sharma BK, Erhan SZ, Biswas A, Wang R, Schuman TP. Oxidation and low temperature stability of polymerized soybean oil-based lubricants. Thermochimica Acta. 2015; **601**:9-16. DOI: 10.1016/j.tca.2014.12.010

[28] Ting C-C, Chen C-C. Viscosity and working efficiency analysis of soybean oil based bio-lubricants. Measurement. 2011;**44**:1337-1341. DOI: 10.1016/j.measurement.2011.04.005

[29] Sharma BK, Perez JM, Erhan SZ. Soybean oil-based lubricants: A search for synergistic antioxidants. Energy &

Fuels. 2007;**21**:2408-2414. DOI: 10.1021/ef0605854

[30] Wang M. Biolubricants and biolubrication [thesis]. Stockholm: KTH Royal Institute of Technology; 2014

[31] Redondo-Cuevasa L, Castellano G, Torrens F, Raikos V. Revealing the relationship between vegetable oil composition and oxidative stability: A multifactorial approach. Journal of Food Composition and Analysis. 2018;**66**: 221-229. DOI: 10.1016/j.jfca.2017.12.027

[32] Fox NJ, Stachowiak GW. Vegetable oil-based lubricants: A review of oxidation. Tribology International. 2007;**40**:1035-1046. DOI: 10.1016/j.triboint.2006.10.001

[33] Erhan SZ, Sharma BK, Perez JM. Oxidation and low temperature stability of vegetable oil-based lubricants. Industrial Crops and Products. 2006;**24**: 292-299. DOI: 10.1016/j.indcrop.2006.06.008

[34] Buosi GM, da Silva ET, Spacino K, Silva LRC, Ferreira BAD, Borsato D. Oxidative stability of biodiesel from soybean oil: Comparison between synthetic and natural antioxidants. Fuel. 2016;**181**:759-764. DOI: 10.1016/j.fuel.2016.05.056

[35] de Sousa LS, de Moura CVR, de Oliveira JE, de Moura EM. Use of natural antioxidants in soybean biodiesel. Fuel. 2014;**134**:420-428. DOI: 10.1016/j.fuel.2014.06.007

[36] Erhan SZ, Adhvaryu A, Sharma BK. Chemically functionalized vegetable oils. In: Rudnick LR, editor. Synthetics, Mineral Oils and Bio-Based Lubricants. Chemistry and Technology. Boca Raton: CRC Press; 2006. pp. 361-388. ISBN: 1-57444-723-8

[37] Dowson D, Higginson GR. Elasto-hydrodynamic Lubrication. Oxford: Pergamon Press; 1977. 250 p. ISBN: 0-08-021303-0

[38] Wan Nik WW, Ani FN, Masjuki HH, Eng Giap SG. Rheology of bio-edible oils according to several rheological models and its potential as hydraulic fluid. Industrial Crops and Products. 2005;**22**:249-255. DOI: 10.1016/j.indcrop.2005.01.005

[39] Yilmaz N. Temperature-dependent viscosity correlations of vegetable oils and biofuel-diesel mixtures. Biomass and Bioenergy. 2011;**35**:2936-2938. DOI: 10.1016/j.biombioe.2011.03.026

[40] Esteban B, Riba J-R, Baquero G, Rius A, Puig R. Temperature dependence and viscosity of vegetable oils. Biomass and Bioenergy. 2012;**42**:164-171. DOI: 10.1016/j.biombioe.2012.03.007

[41] Cristea GC. Tribological characterization of soybean oil additivated with nano materials based on carbon (black carbon, graphite and graphene) [thesis]. Galati: Dunarea de Jos University; 2017

[42] Rudnick LR, editor. Lubricant Additive. Chemistry and Applications. 2nd ed. Boca Raton: CRC Press; 2009. 777 p. ISBN: 978-1-4200-5964-9

[43] Tang Z, Li S. A review of recent developments of friction modifiers for liquid lubricants (2007–present). Current Opinion in Solid State and Materials Science. 2014;**18**:119-139. DOI: 10.1016/j.cossms.2014.02.002

[44] Lubricants and Lubricant Additives [Internet]. 2014. Available from: http://dx.doi.org/10.1787/9789264221154-en [Accessed: 2018-04-14]

[45] Quinchia LA, Delgado MA, Reddyhoff T, Gallegos C, Spikes HA. Tribological studies of potential vegetable oil-based lubricants containing environmentally friendly

viscosity modifiers. Tribology International. 2014;**69**:110-117. DOI: 10.1016/j.triboint.2013.08.016

[46] Friedrich K, Schlarb AK, editors. Tribology of polymeric nanocomposites–Friction and wear of bulk materials and coatings. 1st ed. Oxford: Elsevier; 2008. 568 p. ISBN: 978-0-444-53155-1

[47] Cristea GC, Cazamir D, Dima D, Georgescu C, Deleanu L. Influence of TiO_2 as nano additive in rapeseed oil. In: Proceedings of the 8th International Conference on Advanced Concepts in Mechanical Engineering (ACME). Romania: Iasi; 2018

[48] Fessenbecker A, Roehrs I, Pegnoglou R. Additives for environmentally acceptable lubricants. NLGI Spokesman. 1996;**60**(6):9-25

[49] Shahnazar S, Bagheri S, Hamid SBA. Enhancing lubricant properties by nano particle additives. International Journal of Hydrogen Energy. 2016;**41**:3153-3170. DOI: 10.1016/j.ijhydene.2015.12.040

[50] Zainal NA, Zulkifli NWM, Gulzar M, Masjuki HH. A review on the chemistry, production, and technological potential of bio-based lubricants. Renewable and Sustainable Energy Reviews. 2018;**82**:80-102. DOI: 10.1016/j.rser.2017.09.004

[51] Wu YY, Tsui WC, Liu TC. Experimental analysis of tribological properties of lubricating oils with nanoparticle additives. Wear. 2007;**262**: 819-825. DOI: 10.1016/j.wear.2006.08.021

[52] Wu H, Zhao J, Xia W, Cheng X, He A, Yun JH, et al. A study of the tribological behaviour of TiO_2 nanoadditive water-based lubricants. Tribology International. 2017;**109**: 398-408. DOI: 10.1016/j.triboint.2017.01.013

[53] Wo H, Hu K, Hu XG. Tribological properties of MoS_2 nanoparticles as additive in a machine oil. Tribology. 2004;**24**:33-37

[54] Wu JF, Zhai WS, Jie GF. Preparation and tribological properties of WS_2 nanoparticles modified by trioctylamine. Proceedings of the Institution of Mechanical Engineers, Part J: Journal of Engineering Tribology. 2009;**223**:695-703. DOI: 10.1243/13506501JET494

[55] Chinas-Castillo F, Spikes HA. Mechanism of action of colloidal solid dispersions. Journal of Tribology. 2003; **125**:552-557. DOI: 10.1115/1.1537752

[56] Alves SM, Barros BS, Trajano MF, Ribeiro KSB, Moura E. Tribological behavior of vegetable oil-based lubricants with nanoparticles of oxides in boundary lubrication conditions. Tribology International. 2013;**65**:28-36. DOI: 10.1016/j.triboint.2013.03.027

[57] Gu C, Zhu G, Li L, Tian X, Zhu G. Tribological effects of oxide based nanoparticles in lubricating oils. Journal of Marine Science and Applications. 2009;**8**:71-76. DOI: 10.1007/s11804-009-8008-1

[58] Iliuc I. Tribology of thin layers. 1st ed. New York: North Holland; 1980. 224 p. ISBN: 0-444-99768-7

[59] Yu H, Xu Y, Shi P, B-s X, Wang X, Liu Q. Tribological properties and lubricating mechanisms of Cu nanoparticles in lubricant. Transactions of Nonferrous Metals Society of China. 2008;**18**:636-641. DOI: 10.1016/S1003-6326(08)60111-9

[60] Liu G, Li X, Qin B, Xing D, Guo Y, Fan R. Investigation of the mending effect and mechanism of copper nano-particles on a tribologically stressed surface. Tribology Letters. 2004;**17**: 961-966. DOI: 10.1007/s11249-004-8109-6

[61] Lee K, Hwang Y, Cheong S, Choi Y, Kwon L, Lee J, et al. Understanding the role of nanoparticles in nano-oil lubrication. Tribology Letters. 2009;**35**: 127-131. DOI: 10.1007/s11249-009-9441-7

[62] Tevet O, von-Huth P, Popovitz-Biro R, Rosentsveig R, Wagner HD, Tenne R. Friction mechanism of individual multilayered nanoparticles. Proceedings of the National Academy of Sciences of United States of America. 2011;**108**: 19901-19906. DOI: 10.1073/pnas.1106553108

[63] Akbulut M. Nanoparticle-based lubrication systems. Journal of Powder Metallurgy and Mining. 2012;**1**:e101. DOI: 10.4172/2168-9806.1000e101

[64] Jayadas NH, Nair KP, Ajithkumar G. Tribological evaluation of coconut oil as an environment-friendly lubricant. Tribology International. 2007;**40**: 350-354. DOI: 10.1016/j.triboint.2005.09.021

[65] Lahouij I, Bucholz EW, Vacher B, Sinnott SB, Martin JM, Dassenoy F. Lubrication mechanisms of hollow-core inorganic fullerene-like nanoparticles: Coupling experimental and computational works. Nanotechnology. 2012;**23**:375701. DOI: 10.1088/0957-4484/23/37/375701

[66] Czichos H, Saito T, Smith L, editors. Springer Handbook of Materials Measurement Methods. Berlin: Springer Science-Business Media; 2006. 1208 p. ISBN: 9783540303008

[67] EN ISO 20623:2017. Petroleum and related products. Determination of the extreme-pressure and anti-wear properties of fluids. Four ball method (European conditions)

[68] Tiong CI, Azli Y, Kadir MRA, Syahrullail S. Tribological evaluation of refined, bleached and deodorized palm stearin using four-ball tribotester with different normal loads. Journal of Zhejiang University - Science A (Applied Physics & Engineering). 2012; **13**:633-640. DOI: 10.1631/jzus.A1200021

[69] Spanu C, Ripa M, Stefanescu I, Deleanu L. A comparation of standardized methods for lubrication failure determination. The Annals of "Dunarea de Jos" University of Galati, Fascicle VIII, Tribology. 2007;**XIII**: 99-103

[70] Holmberg K. Quality of reporting empirical results in tribology [Internet]. 2005. Available from: http://virtual.vtt.fi/virtual/proj3/irg/docs/quality_of_reporting_kh190805 _020905.doc [Accessed: 2018-05-12]

[71] Blok H. The flash temperature concept. Wear. 1963;**6**:483-494. DOI: 10.1016/0043-1648(63)90283-7

[72] Marscher WD. A critical evaluation of the Flash-Temperature concept. Journal ASLE Transactions. 1982;**25**: 157-174. DOI: 10.1080/05698198208983077

[73] Deleanu L, Cristea GC, Georgescu C, Suciu C. Texture investigation of scars resulted from lubrication with soybean oil additivated with nanographite. In: Proceedings of the 2nd International Conference on Tribology (TURKEYTRIB'18), 18–20 April 2018, Istanbul, Turkey

[74] Georgescu C, Cristea CG, Dima S, Deleanu L. Evaluating lubricating capacity of vegetal oils using Abbott-Firestone curve. IOP Conference Series: Materials Science and Engineering. 2017;**174**:012057. DOI: 10.1088/1757-899X/174/1/012057

[75] Georgescu C. Use of vegetable oils to obtain ecological lubricants [postdoctoral report]. Galati: Dunarea de Jos University of Galati; 2015

[76] Nanomaterials and related products [Internet]. 2016. Available from: http://www.plasmachem.com/download/PlasmaChem-General_Catalogue_Nanomaterials.pdf [Accessed: 2018-05-12]

[77] Cameron A. Basic Lubrication Theory. 3rd ed. Chichester: Ellis Horwood Ltd.; 1983. 256 p. ISBN: 978-0470275542

[78] Cheenkachorn K. A study of wear properties of different soybean oils. Energy Procedia. 2013;**42**:633-639. DOI: 10.1016/j.egypro.2013.11.065

[79] Zhao X, Yang J, Tao D, Xu X. Tribological study of nitrogen plasma polymerized soybean oil with nitrogen heterocyclic structures. Industrial Crops and Products. 2013;**51**:236-243. DOI: 10.1016/j.indcrop.2013.08.074

[80] Battez AH, Rico JEF, Arias AN, Rodriguez JLV, Rodriguez RC, Fernandez JMD. The tribological behaviour of ZnO nanoparticles as an additive to PAO6. Wear. 2006;**261**: 256-263. DOI: 10.1016/j.wear.2005.10.001

[81] Zulkifli NWM, Kalam MA, Masjuki HH, Yunus R. Experimental analysis of tribological properties of biolubricant with nanoparticle additive. Procedia Engineering. 2013;**68**:152-157. DOI: 10.1016/j.proeng.2013.12.161

[82] Cristea GC, Dima C, Georgescu C, Dima D, Deleanu L, Solea LC. Evaluating lubrication capability of soybean oil with nano carbon additive. Tribology in Industry. 2018;**40**(1): 66-72. DOI: 10.24874/ti.2018.40.01.05

[83] Cristea GC, Dima C, Dima D, Georgescu C, Deleanu L. Nano graphite as additive in soybean oil. MATEC Web of Conferences. 2017;**112**:04023. DOI: 10.1051/matecconf/201711204023

Chapter 6

Row Spacing and Seeding Rate Effects on Soybean Seed Yield

Matthew Schutte and Thandiwe Nleya

Abstract

Soybean growers in the northern latitudes of the United States plant the crop in a wide range of row spacings although there has been a shift toward wider rows (>50 cm) in some Upper Midwest states in the last 5 years. The objective of this study was to evaluate the impact of row spacing and seeding rate on the performance of soybean and to determine whether these management practices interact to influence soybean yield. A row spacing study was conducted at Aberdeen and Beresford, South Dakota, USA, in 2014 and 2015. The study had two row spacings (19 and 76 cm), four seeding rates (247,000, 333,500, 420,000, and 506,500 seeds ha^{-1}), and two soybean varieties at each location. Soybean had greater stand establishment in 19 cm rows (6–10% higher) compared with 76 cm rows. Soybean in 19 cm rows yielded 0.8–10% more than in 76 cm rows depending on the location or year. Seed yield increased with increasing seeding rate with the highest seeding rate of 506,000 seeds ha^{-1} yielding greatest. The increase in seed yield due to the increase in seeding rate ranged from 3 to 7%. At each location, the longer duration soybean variety yielded higher than the shorter duration variety.

Keywords: soybean, *Glycine max*, row spacing, seeding rate, seed yield

1. Introduction

Soybean (*Glycine max*) is the second most planted crop after corn worldwide and is the second most important source of crop revenue in South Dakota [1]. Research conducted in the Upper Midwest of the United States documents a consistent yield advantage, in the range of 134–604 kg ha^{-1}, for soybean grown in narrow row spacings (<50 cm) when compared to those grown at wider row spacings (50–76 cm) [2–4]. Another research, however, showed no yield advantage to narrow row spacing [5]. Cox and Cherney [6] reported that soybean drilled in 19 cm rows yielded 7% more than soybeans planted with a row crop planter in 38 cm rows and 17% more than soybean planted in 76 cm rows in Northeastern United States. Even with these reports of yield advantage or no yield difference, 69% of soybean growers in South Dakota, 54% in Nebraska, and 49% in Iowa grow soybean in 76 cm row spacing or wider [1].

Lee [7] reported that in Central and Southern United States row spacing studies usually found no increase in yield in narrow rows over wider rows. This was confirmed by Thompson et al. [8] who reported that yield responses to narrow row spacing in the Mid-South United States were inconsistent and mainly influenced by weather. The increase in yield from narrow row spacings in the Northern United

IntechOpen

States has been attributed to a shorter growing season meaning soybean has limited time to reach maximum radiation interception prior to flowering. Narrow rows therefore increase radiation interception during the critical periods for grain set resulting in earlier canopy closure and less light being usable for weeds if initial weed control is satisfactory [9–12]. Along with higher rate of light interception, less evapotranspiration was reported in narrow rows due to faster canopy closure and thus resulted in a higher water-use efficiency [13]. However, in years of drought stress, narrow rows can deplete soil water sooner by increased vegetative growth and result in insufficient soil water availability during reproductive stages and therefore no yield advantage over wider rows [2, 14].

Some studies have reported row spacing × seeding rate interactions with soybean yielding greater with higher seeding rates and narrow rows when compared to wide rows [3, 6, 15, 16]. Cox et al. [3] reported a greater profit of US$30 ha^{-1} with a seeding rate of 420,000 seeds ha^{-1} in 19 cm rows compared to 321,000 seeds ha^{-1} in 76 cm rows due to yield increase outweighing seed costs. Other studies have reported similar optimum seeding rates between narrow and wide rows and therefore no interaction between row spacing and seeding rate [17–19]. Ricks et al. [20] reported that the optimum seeding rates for South Dakota typically range between 355,000 seeds ha^{-1} and 381,000 seeds ha^{-1}. However, they also reported that higher yields have been reported with seeding rates greater than 406,000 seeds ha^{-1}.

Carpenter and Board [21] reported that soybean plants compensate for space in the canopy by adding branches, and they found no yield response to increased seeding rates. This was supported by Cox and Cherney [6] who found that not only did soybean plants compensate with biomass, pods, and seeds per plant at lower seeding rates but also found that soybean compensated for wider rows (>38 cm) as well. They also found that though soybean plants do compensate for both lower seeding rates and wider rows, they were less efficient at compensating for wider rows than for lower seeding rates, meaning that row spacing had a greater effect on yield than seeding rate. Wiatrak and Chen [22] found that increasing seeding rate may improve soybean growth at early vegetative stages, which in turn can result in increase in yield. However, they found that seeding rates above 272,000 seeds ha^{-1} did not follow this trend and did not increase vegetative growth.

White mold (also called *Sclerotinia* stem rot), a disease caused by the fungus *Sclerotinia sclerotiorum*, is a yield-limiting soybean disease in North Central United States. Management practices such as narrow row spacing, increased plant populations, early planting dates, and high-soil fertility can increase soybean yields but have the unintended consequence of increasing white mold development within the soybean canopy [23, 24]. While fungicides are available to control white mold, complete control of the disease using only chemical management is usually not possible [24]. Thus, in addition to fungicides, management strategies for controlling white mold in soybean include cultivars selection and management practices to reduce canopy density [24, 25]. Planting in wide row spacings or at lower plant populations delays canopy closer, reduces canopy density, and thus prevents favorable conditions for white mold development [24, 26].

With increase in soybean planted in wider rows (50–76 cm) in South Dakota and neighboring states in the Upper Midwest, there is a need to evaluate the value of this practice especially with recent research results suggesting that narrow rows have an advantage or at least yield the same as wider rows in the Upper Midwest. The objectives of this study were to (i) determine the effect of row spacing and seeding rate on soybean yield and (ii) measure the interactions between the two management practices.

2. Materials and methods

The study was conducted at two locations, Southeast Research Farm, Beresford, South Dakota (SD) (43.052548°N, 96.904135°W), and Aberdeen, SD (45.464698° N, 98.486483°W) in 2014 and 2015. At Beresford, the soil textural classification was Egan-Clarno-Chancellor complex, fine silty, and fine loam [27]. At Aberdeen, the soil textural classification was Great Bend fine silty, mixed, superactive, and frigid calcic Hapludolls [28]. The experimental fields were plowed in the fall and cultivated twice in the spring before planting soybean. The soybean was grown under dryland conditions. The total rainfall and mean air temperature for each growing season are shown in **Table 1**.

The experimental design was a randomized complete block in a split-plot arrangement, with four replications. The main plots were two row spacings. Subplot treatments were four seeding rates of 247,000, 333,500, 420,000, and 506,500 viable seeds ha^{-1} and two soybean varieties arranged in a factorial design. The two row spacings were 19 and 76 cm rows. The soybean varieties were different at each location based on maturity grouping ideal for the area and were also slightly different in resistance to white mold. At the Aberdeen location, the varieties were 0906R2 and 1108R2 and at Beresford were 2306R2 and 2408R2 (Channel, St. Louis, MO). At each specific location, varieties 0906R2 and 2306R2 were of shorter duration than 1108R2 and 2408R2. The rating for white mold were 3 for 0906R2, 4 for 1108R2, 3 for 2306R2, and 6 for 2408R2 on a scale of 1–9 (1 resistant and 9 susceptible) [29].

In 2014, the planting dates were June 9 and May 28 at Aberdeen and Beresford, respectively. In 2015, the planting dates were June 9 at Aberdeen and June 10 at Beresford. For the 76 cm row spacing, soybean was planted in four rows that was 6.4 m long and trimmed back to 5.5 m when they reached the V3 stage. The center

Average monthly temperature (°C)								
Location	**Year**	**May**	**June**	**July**	**August**	**September**	**October**	**Average**
Aberdeen	2014	12.89	17.53	19.61	19.58	15.33	9.14	15.68
Aberdeen	2015	12.94	20.56	22.58	20.42	18.39	10.44	17.56
30-year average		13.55	18.65	21.80	20.56	14.95	7.29	16.1
Beresford	2014	15.31	20.19	20.50	21.14	16.47	10.31	17.32
Beresford	2015	14.58	20.83	22.14	20.22	19.61	11.42	18.13
30-year average		15.03	20.53	22.81	21.56	16.58	9.41	16.2
Monthly rainfall (mm)								**Total**
Aberdeen	2014	55.37	84.07	17.78	157.23	25.40	6.60	346.46
Aberdeen	2015	162.31	53.34	103.12	74.68	9.40	41.66	444.50
30-year average		78.99	93.98	75.95	61.72	55.63	50.55	416.60
Beresford	2014	62.99	342.90	27.18	75.18	61.47	34.54	604.27
Beresford	2015	89.66	90.42	150.11	179.07	92.46	26.42	628.14
30-year average		92.46	110.74	83.31	72.39	74.42	54.61	487.90

Source: High Plains Regional Climate Center, University of Nebraska, http://xmacis.rcc-acis.org/#, last accessed 6/13/2018.

Table 1.
Monthly average air temperature and rainfall at Aberdeen and Beresford, SD, for 2014 and 2015.

two rows were harvested for yield data, while the outer two rows were buffers. For the 19 cm row spacing, soybean was planted in 16 rows that is 6.5 m long and trimmed back to 5.5 m at V3 stage. The eight center rows were harvested for yield data with eight buffer rows on either side. The data collected included the number of plants ha^{-1} at the V4 growth stage determined by counting the number of plants in the middle two rows for the 76 cm row spacing and eight rows for the 19 cm row spacing and converting to plants ha^{-1}. Seed yield was determined by harvesting two center rows (76 cm spacing) and eight center rows (19 cm spacing) with a small-plot combine (Massey Ferguson 8XP, Duluth, Georgia, USA). Seed subsamples from each plot were taken to determine moisture, protein, and oil content. Seed moisture was determined by weighing seed samples before drying at 60°C for 48 hours and reweighing the samples after drying to adjust seed moisture to 13% or 130 g kg^{-1}. Seed protein and seed oil were determined using a near-infrared transmittance (NIT) spectroscopy (Infratec 1229 Grain Analyzer, Foss Tecator AB).

Weeds were managed with a preemergent herbicide application of S-metolachlor (Dual II) (Bayer CropScience, Research Triangle Park, NC) and two in-season application of glyphosate (PowerMax) (Monsanto Company, St. Louis, MO). The insecticide Baythroid [cyano(4-fluoro-3-phenoxyphenyl)methyl-3-(2,2-dichloro-ethenyl)-2,2-dimethyl-cyclopropanecarboxylate] (Bayer CropScience, Research Triangle Park, NC) was applied when soybean aphids (*Aphis glycines*) reached economic thresholds.

Data were analyzed using PROC MIXED of SAS (SAS Research Institute, NC). Years and blocks were treated as random, and all other effects were considered fixed. Levene's test was used to test for the homogeneity of variance. After combined analysis revealed interactions between location and year, the data were split by year and then by location to analyze the significant interactions between row spacing, variety, and seeding rate within each location. Mean separation was performed using Fisher's protected LSD (0.05).

3. Results and discussion

3.1 Climate and weather

Average temperatures were slightly warmer at Beresford compared to Aberdeen, although in 2015, September was much warmer compared to 2014 at both locations (**Table 1**). Rainfall amounts and timing varied considerably for each location and each year. Aberdeen was drier (70.1 mm less rain) than long-term average in 2014 and wetter (28.1 mm more) than long-term average in 2015. Beresford was wetter than long-term average in both years with June 2014 receiving 132.1 mm more rain than average. The warmer and wetter conditions at Beresford in both years were conducive to overall better soybean growth and yield when compared to Aberdeen.

3.2 Established plant population

In 2014, the effects of row spacing on number of plants ha^{-1} and percent stand establishment (relative to seeding rate) were significant (<0.001) at both locations, while in 2015, row spacing effects were significant for the two traits ($P = 0.02$ and 0.01, respectively) only at Aberdeen (**Table 2**). Overall, plant establishment was greater in narrow rows compared with wide rows. On average, the difference in stand establishment between the two row spacings was greater at the Aberdeen location (10% points) compared to Beresford (6% points). Greater stand

	2014				2015			
	Aberdeen		Beresford		Aberdeen		Beresford	
	Plants (ha^{-1})	Percentage (%) stand	Plants (ha^{-1})	Percentage (%) stand	Plants (ha^{-1})	Percentage (%) stand	Plant (ha^{-1})	Percentage (%) stand
Row spacing (S) (cm)								
19	352,975a*	96.7a	315,660a	85.1a	324,032a	86.2a	316,557	85.3
76	279,071b	75.7b	286,695b	76.6b	288,638b	77.1b	307,811	82.7
Seeding rate (RS) (seeds ha^{-1})								
247,000	230,821d	93.4 a	208,247d	84.4a	204,585d	82.8	220,431d	89.2a
333,500	288,003c	86.3b	281,575c	84.3a	276,940c	83.0	290,395c	87.0ab
420,000	345,634b	82.8c	334,048b	79.5b	345,335b	82.2	346,755b	82.6b
506,500	419,634a	82.2c	380,840a	75.2b	398,480a	76.7	391,155a	77.2c
Variety (V)#								
0906R2/2306R2	323,733a	87.1a	301,981	80.9	302,467	80.5	306,690	82.3
1108R2/2408R2	318,313b	85.3b	300,374	80.7	310,203	82.2	317,678	85.6
	Analysis of variance							
S	<0.001	<0.001	0.025	0.009	0.020	0.016	0.075	0.097
SR	<0.001	<0.001	<0.001	<0.001	<0.001	0.091	<0.001	<0.001
S × SR	<0.001	<0.001	0.316	0.069	0.036	0.604	0.444	0.521
V	0.048	0.028	0.811	0.850	0.141	0.091	0.079	0.053
V × S	0.748	0.688	0.539	0.560	0.086	0.062	0.243	0.232
V × SR	0.524	0.172	0.992	0.993	0.424	0.166	0.181	0.197
V × SR × S	0.758	0.722	0.451	0.538	0.946	0.928	0.631	0.512

*Within each column and each treatment, means followed by the same letter are not significantly different (P 0.05).
#Soybean varieties 0906R2 and 1108R2 were grown at Aberdeen and 2306R2 and 2408R2 at Beresford.

Table 2.
Established plant population and percentage (%) established plants (relative to seeding rate) at Aberdeen and Beresford locations, SD, in 2014 and 2015.

establishment in narrow rows has been observed by others in the Upper Midwest [2, 16]. As expected increasing seeding rate increased the number of established plant ha^{-1} at both locations and in both years. Percent established plants relative to the target population, on the other hand, decreased significantly as the seeding rate increased, and this was true in three of the four location-years. The rate of decrease in percent established plants was variable among location-years ranging from a high 12% drop between the lowest and the highest seeding rates at Beresford in 2015 to the lowest drop of 6.1% at Aberdeen in 2015. The reason for this is not clear, but Bruns [30] also reported a decrease in percent established plants with increasing seeding rate. However, it is generally accepted that under optimal conditions, stand establishment is about 80% of the seeding rate [30, 31]. In this study we achieved 80% stand establishment for all seeding rates except for the highest seeding rate of 506,500 at Beresford in 2014 and 2015 and in Aberdeen in 2015.

The row spacing × seeding rate interaction effects were significant at Aberdeen in both years (**Tables 2 and 3**). The interaction was due to the fact that the decrease in the number of established plants or percent stand establishment with increase in seeding rate was lower for the 19 cm row spacing when compared to the wider row spacing in both years (4.7% vs. 17.2% in 2014; 2.4% vs. 5.9% in 2015).

3.3 Seed yield

Row spacing, seeding rate, and variety effects on seed yield were significant in both years at Aberdeen and in 2015 at Beresford (**Tables 4** and **5**). In 2014, only seeding rate significantly affected seed yield at Beresford. In all four location-years, the narrow row spacing of 19 cm outyielded the wider row spacing of 76 cm with the yield advantage ranging from 37 to 424 kg ha^{-1} or 0.8 to 10%. Our results agree with earlier finding by other researchers in the Upper Midwest [2–4]. The advantage of narrow rows in the Northern United States is attributed to a shorter growing season and related canopy development and light interception. Narrow rows speed the rate of canopy closure and hence increase light interception [11, 12]. Earlier canopy closure means less moisture loss through evapotranspiration and results in higher water-use efficiency [13]. However, it is important to note that the advantage of narrow rows can diminish under moisture stress. Soybean plants grown in

		2014		2015	
Row spacing (S) (cm)	**Seeding rate (SR) (seeds ha^{-1})**	**Plant (ha^{-1})**	**Percentage (%) stand**	**Plant (ha^{-1})**	**Percentage (%) stand**
19	247,000	246,368	99.7	215,273	87.1
	333,500	323,209	96.9	288,825	86.6
	420,000	397,359	94.6	362,975	86.4
	506,500	484,963	95.7	429,052	84.7
76	247,000	215,273	87.1	193,896	78.5
	333,500	252,786	75.8	265,055	79.4
	420,000	293,908	69.9	327,694	77.9
	506,500	354,304	69.9	367,908	72.6
SE		3759	1.08	7306	1.9

Table 3.
Interaction of row spacing and seeding rate for established plants ha^{-1} and percentage (%) stand establishment at Aberdeen, SD, in 2014 and 2015.

	Aberdeen			Beresford		
Row spacing (S) (cm)	**Yield (kg ha^{-1})**	**Seed protein (g kg^{-1})**	**Seed oil (g kg^{-1})**	**Yield (kg ha^{-1})**	**Seed protein (g kg^{-1})**	**Seed oil (g kg^{-1})**
19	4189a*	336.1	180.2	4765	347.9a	178.0
76	3765b	321.7	179.9	4728	344.3b	179.0
Seeding rate (SR) (seeds ha^{-1})						
247,000	3863b	307.8b	180.9	4542c	343.5c	179.5a
333,500	3964b	333.2ab	180.2	4743b	344.4bc	179.2ab
420,000	3986ab	336.1a	179.6	4832ab	346.6b	178.1bc
506,500	4095a	336.4a	179.6	4868a	350.0a	177.4c
Variety (V)#						
0906R2/ 2306R2	3888b	327.4	179.2b	4765	344.0b	178.8
1108R2/ 2408R2	4067a	329.3	180.9a	4727	348.2a	178.3
	Analysis of variance (P > F)					
S	<0.001	0.187	0.549	0.566	<0.001	0.121
SR	0.007	0.113	0.199	<0.001	<0.001	0.004
S × SR	0.853	0.470	0.971	0.192	0.228	0.131
V	<0.001	0.841	0.001	0.386	<0.001	0.258
V × S	0.024	0.408	0.098	0.056	0.699	0.887
V × SR	0.195	0.428	0.147	0.249	0.143	0.608
V × S × SR	0.823	0.461	0.777	0.639	0.705	0.393

Within each column and each treatment, means followed by the same letter are not significantly different (P 0.05).
#Soybean varieties 0906R2 and 1108R2 were grown at Aberdeen and 2306R2 and 2408R2 at Beresford.

Table 4.
Seed yield, seed protein concentration, and seed oil concentration of soybean as influenced by row spacing, seeding rate, and variety at two locations in South Dakota in 2014.

narrow rows can deplete soil water early in the growing season resulting in insufficient available water during the reproduction stages of growth [14, 20].

Seeding rate effects for seed yield were significant for both years and locations (**Tables 4** and **5**). In all four location-years, the top seeding rate of 506,500 seeds ha^{-1} yielded significantly higher than the other three seeding rates, while the three lower seeding rates of 247,000, 333,500 and 420,000 had similar yields at Aberdeen in 2014 and 2015 and at Beresford in 2015. Carpenter and Board [21], Cox et al. [32], and Thompson et al. [8] reported no yield response of soybean to seeding rate and attributed this to the fact that soybean compensates for space in the canopy by adding more branches. Similarly, Cox and Cherney [6] reported that soybean compensated with more biomass, pods, and seed plant^{-1} at lower seeding rates. On the other hand, other researchers have reported that increasing seeding rate can result in greater yield [22, 31]. While the present study supports the later research findings, it is important to note that the seed yield increase observed in this study due to seeding rate was very low ranging from 3 to 7%. This supports the reported [6] compensatory power of soybean plants at lower seeding rates.

Row spacing × seeding rate interaction for seed yield was significant only at one location-year (Beresford, 2015). The interaction was due to the fact that the narrow

	Aberdeen			Beresford		
Row spacing (S) (cm)	Yield (kg ha^{-1})	Seed protein (g kg^{-1})	Seed oil (g kg^{-1})	Yield (kg ha^{-1})	Seed protein (g kg^{-1})	Seed oil (g kg^{-1})
19	4174a*	325.8	195.3b	4521a	331.4a	195.0
76	4018b	326.7	198.7a	4325b	328.4b	195.8
Seeding rate (SR) (seeds ha^{-1})						
247,000	4042b	323.2	197.7	4390b	329.4	195.4
333,500	4068b	328.3	196.8	4394b	330.1	195.8
420,000	4087b	325.4	197.1	4395b	329.8	195.5
506,500	4185a	326.2	196.6	4510a	330.3	194.9
Variety (V)#						
0906R2/ 2306R2	4058b	322.7b	197.1	4319b	328.1	195.7
1108R2/ 2408R2	4133a	328.8a	196.9	4526a	330.7	195.1
	Analysis of variance (P > F)					
S	<0.001	0.956	0.041	0.003	0.021	0.372
SR	0.003	0.097	0.605	0.008	0.965	0.774
S × SR	0.155	0.621	0.892	0.029	0.089	0.915
V	0.008	<0.001	0.839	<0.001	0.282	0.335
V × S	0.895	0.018	0.160	0.269	0.069	0.771
V × SR	0.004	0.675	0.008	<0.001	0.384	0.065
V × S × SR	0.038	0.682	0.221	0.487	0.948	0.154

**Within each column and each treatment, means followed by the same letter are not significantly different (P 0.05).*
#Soybean varieties 0906R2 and 1108R2 were grown at Aberdeen and 2306R2 and 2408R2 at Beresford.

Table 5.
Seed yield, seed protein concentration, and seed oil concentration of soybean as influenced by row spacing, seeding rate, and variety at two locations in South Dakota in 2015.

row spacing of 19 cm yielded significantly higher than the wider row spacing (76 cm) only at higher seeding rates of 420,000 (yield 5% higher) and 506,500 (yield 7% higher) (data not presented). Previous research results on row spacing × seeding rate interactions are in dispute with some researchers [3, 6] reporting row spacing × seeding rate interactions and soybean yielding greater at higher seeding rates and narrow row spacing as reported at Beresford in 2015. Other researchers have reported similar optimum seeding rates for both narrow and wider rows [8, 18, 19]. The current results are more in agreement with the later reports as 3 of 4 location-years did not show significant row spacing × seeding rate interaction.

Variety effects for seed yield were significant at Aberdeen in 2014 and 2015 and at Beresford in 2015. The varieties were chosen based on adaptation to the region but also were different in white mold ratings. At each location, the longer duration variety had a higher white mold rating (less resistant) than the shorter duration variety. In both years and in all instances, where varietal effects were significant, the longer duration variety was the higher yielding of the two. However, the difference was not considered to be related to white mold since white mold scouting showed little to no white mold infection in both years and locations. Instead, the yield difference is attributable to season length and the longer duration variety maximizing yield due to extra growing days. This was supported by the fact that variety × row spacing interaction effects on seed yield were significant only in one location-year (Aberdeen, 2014). Even then, the interaction was due to the longer duration variety (1108R2) yielding significantly higher than the shorter duration variety (0906R2) (3906 vs. 3624 kg ha^{-1}) when seeded in 76 cm row spacings, but the two varieties yielding the same (4227 vs. 4151 kg ha^{-1}) when seeded in 19 cm rows. White mold, if present, would be a bigger problem under narrow rows due to high humidity under a dense canopy [4, 24]. The fact that the row spacing × variety interaction was observed in only 1 year and under wider rows further confirms that the yield advantage of long duration varieties was related to season length.

Variety × seeding rate effects on seed yield were significant at both locations in 2015 (**Table 5**). The interactions are presented in **Table 6**. At Aberdeen the interaction was due to the fact that the longer duration variety showed an increase in seed yield with increasing seeding rate with the best yield obtained at a seeding rate of 506,500 seeds ha^{-1}. For the short duration variety, however, trends were different with the lowest seeding rate of 247,000 seeds ha^{-1} yield the same as the highest seed rate (**Table 6**). At Beresford, the variety × row spacing interaction was, again, due to inconsistent performance of varieties at different seeding rates with the longer duration variety yielding highest at the lowest seeding rate. These results are not surprising as soybean plants respond to environmental conditions and can compensate for lower plant populations by producing more branches [32].

3.4 Seed protein and seed oil concentration

Row spacing, seeding rate, and variety effects for seed protein concentration were significant at Beresford in 2014 (**Table 4**). Seed from narrow rows had higher protein than from wider rows, while protein concentration increased with increasing seeding rate, and the longer duration soybean variety had higher seed protein

	Yield (kg ha^{-1})	
Seeding rate (seeds ha^{-1})	**0906R2**	**1108R2**
	Aberdeen (2015)	
247,000	4103a*	4034b
333,500	3985b	4099ab
420,000	3980b	4196a
506,500	4166a	4204a
	Beresford (2015)	
	2306R2	2408R2
247,00	4178b	4602a
333,500	4326a	4464b
420,000	4352a	4439b
506,500	4420a	4601a

**Within each column and year, means followed by the same letter are not significantly different (P 0.05).*

Table 6.
Seed yield of soybean as influenced by seeding rate and variety at two locations in South Dakota in 2015.

than the shorter duration variety. In 2015, variety × row spacing effects were significant for protein at Aberdeen, while row spacing effects were significant at Beresford (**Tables 4** and **5**). The longer duration variety had higher seed protein at Aberdeen in 2015, while the narrow row spacing, again, had higher seed protein than the wider rows at Beresford in 2015. In 2014, variety effects were significant for seed oil concentration at Aberdeen, while seeding rate effects were significant at Beresford. The longer duration variety, 1108R2, had higher seed oil concentration than the shorter duration variety, 180.9 and 179.2 g kg^{-1}, respectively. At Beresford, seed oil concentration decreased with increasing seeding rate with the highest seeding rate of 506,500 seed ha^{-1} having 2.1 g kg^{-1} lower oil concentration than the lowest seeding rate. In 2015, row spacing and variety × seeding rate effects for seed oil concentration were significant at Aberdeen (**Table 5**). The wider row spacing had significantly higher seed oil concentration than the narrow row spacing (198.7 vs. 195.3 g kg^{-1}). There were no clear trends to explain the variety × seeding rate interaction for seed oil concentration rather than that oil concentrations for both varieties were inconsistent from one seeding rate to the other. Research results on the effects of row spacing or seeding rate on protein content and seed oil concentration are not readily available. One consistent relationship, among studies, has been a negative correlation between seed protein and seed oil concentration. This negative correlation can be attributed to various genetic and environmental factors [33]. One possible explanation for the inconsistent relationship between row spacing and seeding rate and grain quality could be explained by water availability during seed filling. Rotundo and Westgate [34] found that water stress during seed filling (R5–R7) reduced protein and oil accumulation in soybean. Accounting for differences in water availability during seed filling and season could explain the major differences in research results for the row spacing and seeding rate studies. For example, longer duration varieties have prolonged seed maturation period resulting in greater oil or protein accumulation. Wider rows may preserve soil moisture making soil moisture conditions more favorable during the seed filling period and therefore greater oil concentration in the seed.

4. Conclusions

A considerable number of growers in the Upper Midwest continue to grow soybean in wide row spacings (50–76 cm). Results from the present study and others indicate that soybean planted in narrow rows of 19 cm have higher yield potential when compared to soybean planted in wider rows. Soybean yield responded to seeding rate with maximum yield obtained at a seeding rate of 506,500 seeds ha^{-1} with no significant interaction between row spacing and seeding rate. In terms of soybean variety, the longer duration variety at each location had higher yield. Although the current results indicate that the best soybean yield can be obtained when the crop is seeded in row spacings of 19 cm at seeding rates of 506,500 seeds ha^{-1}, it must be noted that management choices for growers are influenced by a number of factors. In addition to yield potential, growers consider equipment costs associated with changing row spacings and disease and lodging problems associated with narrow rows or high seeding rates. And because of high costs of soybean seed, economic optimum seeding rates are usually less than seeding rates that result in highest yields. However, it is important that growers in the Upper Midwest consider seeding soybean in narrower rows as the current results and many others show that soybean planted with such row spacings have higher yield potential than soybean planted in wider rows.

Acknowledgements

The project was funded by the South Dakota Soybean Research and Promotion Council and South Dakota Agricultural Experiment Station. We thank Kevin Kirby, Shawn Hawks, and Christopher Owusu for providing technical assistance.

Author details

Matthew Schutte[2] and Thandiwe Nleya[1*]

1 Agronomy, Horticulture, and Plant Science Department, South Dakota State University, Brookings, SD, USA

2 Business Development Representative, Bioverse Natural Solutions, Rushmore, MN, USA

*Address all correspondence to: thandiwe.nleya@sdstate.edu

References

[1] USDA-NASS, 2015. Crop Production Survey. https://www.nass.usda.gov/Statistics_by_State/South_Dakota/index.php.

[2] De Bruin JL, Pedersen P. Effect of row spacing and seeding rate on soybean yield. Agronomy Journal. 2008; **100**:704-710

[3] Cox WJ, Orlowski J, Ditommaso A, Knoblauch W. Planting soybean with a grain drill inconsistently increases yield and profit. Agronomy Journal. 2012;**104**: 1065-1073

[4] Lambert DM, Lowenberg-DeBoer J. Economic analysis of row spacing for corn and soybean. Agronomy Journal. 2003;**95**:564-573

[5] Pedersen P, Lauer JG. Corn and soybean response to rotation sequence, row spacing, and tillage system. Agronomy Journal. 2003;**95**:965-971

[6] Cox WJ, Cherney JH. Growth and yield responses of soybean to row spacing and seeding rates. Agronomy Journal. 2011;**103**:123-128. DOI: 10.2134/agronj2010.0316

[7] Lee CD. Reducing row spacing to increase yield: Why it doesn't always work. Crop Management. St. Paul, MN: Plant Management Network; 2006. Available at: www.plantmanagementnetwork.org/cm/. DOI: 10.1094/CM-2006-0227-04-RV

[8] Thompson NM, Larson JA, Lambert DM, Roberts RK, Mengistu A, Bellaloui N, Walker ER. Mid-south soybean yield and net return as affected by plant population and row spacing. Agronomy Journal. 2015;**107**:979-989

[9] Burnside OC, Colville WL. Soybean and weed yields as affected by irrigation, row spacing, tillage, and Amiben. Weeds. 1964;**12**:109-112

[10] Dougherty CT. The influence of planting date, row-spacing, and herbicides on the yield of soybeans in Canterbury. New Zealand Journal of Agricultural Research. 1969;**12**:703-726

[11] Howe OW III, Oliver LR. Influence of soybean (*Glycine max*) row-spacing on pitted morning glory (*Ipomoea lacunosa*) interference. Weed Science. 1987;**35**:185-193

[12] Andrade FH, Calvino P, Cirilo A, Barbieri P. Yield responses to narrow rows depends on increased radiation interception. Agronomy Journal. 2002; **94**:975-980

[13] Alessi J, Power JF. Influence of moisture, plant population, and nitrogen on dryland corn in the Northern Plains. Agronomy Journal. 1965;**56**:611-612

[14] Reicosky DC, Kaspar TC, Taylor HM. Diurnal relationship between evapotranspiration and leaf water potential of field-grown soybeans. Agronomy Journal. 1982;**74**:667-673

[15] Weber CR, Shibles RM, Byth DF. Effect of plant population and row spacing on soybean development and production. Agronomy Journal. 1966;**58**: 99-102

[16] Oplinger ES, Philbrook BD. Soybean planting date, row width, and seeding rate response in three tillage systems. Journal of Production Agriculture. 1992; **5**:94-99

[17] Beurelein JE. Yield of indeterminate and determinate semidwarf soybeans for several planting dates, row spacings, and seeding rates. Journal of Production Agriculture. 1988;**1**:300-303

[18] Ablett GR, Beversdorf WD, Dirks VA. Row width and seeding rate

performance of indeterminate, semideterminate, and determinate soybean. Journal of Production Agriculture. 1991;**4**:391-395

[19] Kratochvil RJ, Pearce JT, Harrison MR Jr. Row-spacing and seeding rate effects on glyphosate-resistant soybean for mid-Atlantic production systems. Agronomy Journal. 2004;**96**:1029-1038. DOI: 10.2134/agronj2004.1029

[20] Ricks D, Christensen R, Carlson CG. Growing 100-bushel soybeans. In: Clay DE, Carlson CG, Clay SA, Wagner L, Deneke D, Hay C, editors. iGrow Soybean: Best Management Practices for Soybean Production. SDSU Extension, Brookings, SD: South Dakota State University; 2013

[21] Carpenter AC, Board JE. Branch yield components controlling soybean yield stability across plant populations. Crop Science. 1997;**37**:885-891

[22] Chen G, Wiatrak P. Seeding rate effects on soybean maturity group IV-VIII for the Southeastern production system: I. Vegetation indices. Agronomy Journal. 2011;**103**:32-37

[23] Pennypacker BW, Risius ML. Environmental sensitivity of soybean cultivar response to *Sclerotinia sclerotiorum*. Phytopathology. 1999;**89**: 618-622

[24] Crop Protection Network. White mold. Soybean Disease Management, CPN-1005; 2015. WhiteMold_CPN1005_2015.Pdf

[25] Hall R, Nasser LCB. Practice and precept in cultural management of bean diseases. Canadian Journal of Plant Pathology. 1996;**18**:176-185

[26] Buzzell RI, Welacky TW, Anderson TR. Soybean cultivar reaction and row width effect on Sclerotinia stem rot. Canadian Journal of Plant Science. 1993; **73**:1169-1175

[27] USDA-NRCS. Web Soil Survey. Egan-Clarno-Chancellor complex. National Cooperative Soil Survey; 2017. http//:websoilsurvey.sc.egov.usda.gov/App/WebSoilSurvey.aspx [Accessed: 25 July 2018]

[28] USDA-NRCS. Web Soil Survey. Great Bend-Beotia. National Cooperative Soil Survey; 2017. http//:websoilsurvey.sc.egov.usda.gov/App/WebSoilSurvey.aspx [Accessed: 25 July 2018]

[29] Channel Seeds. https//:www.channel.com/seedfinder/Pages/default.aspx [Accessed: 25 July 2018]

[30] Bruns HA. Planting date, rate and twin-row vs. single-row soybean in the mid-south. Agronomy Journal. 2011; **103**:1308-1313

[31] Heartherly LG. Early Soybean Production System (ESPS). In: Heartherly LG, Hodges HF, editors. Soybean Production in the mid-south. Boca Raton, FL: CRC Press; 1999. pp. 103-118

[32] Cox WJ, Cherney JH, Shields E. Soybeans compensate at low seeding rates but not at high thinning rates. Agronomy Journal. 2010;**102**:1238-1243

[33] Watanabe I, Nagasawa T. Appearance and chemical-composition of soybean seeds in germplasm collection of Japan. 2. Correlation among protein, lipid and carbohydrate percentage. Japanese Journal of Crop Science. 1990;**59**:661-666. DOI: 10.1626/jcs.59.661

[34] Rotundo JL, Westgate ME. Meta-analysis of environmental effects on soybean seed composition. Field Crops Research. 2009;**110**:147-156. DOI: 10.1016/j.fcr.2008.07.012

Chapter 7

Beneficial Plant Microbe Interactions and Their Effect on Nutrient Uptake, Yield, and Stress Resistance of Soybeans

Arjun Kafle, Kevin Garcia, Vincent Peta, Jaya Yakha, Alex Soupir and Heike Bücking

Abstract

Plants are meta-organisms that are associated with complex microbiomes. Many of the microorganisms that reside on plant surfaces (epiphytes) or within plant tissues (endophytes) do not cause any plant diseases but often contribute significantly to the nutrient supply of their host plant and can help the plant to overcome a variety of biotic or abiotic stresses. The yield potential of any plant depends not only on successful plant traits that improve, for example, the adaptation to low input conditions or other stressful environments but also on the plant microbiome and its potential to promote plant growth under these conditions. There is a growing interest to unravel the mechanisms underlying these beneficial plant microbe interactions because the activities of these microbial communities are of critical importance for plant growth under abiotic and biotic stresses and could lead to the development of novel strategies to improve yields and stress resistances of agronomically important crops. In this chapter, we summarize our current understanding of the beneficial interactions of soybean plants with arbuscular mycorrhizal fungi, nitrogen-fixing rhizobia, and fungal and bacterial endophytes and identify major knowledge gaps that need to be filled to use beneficial microbes to their full potential.

Keywords: arbuscular mycorrhizal symbiosis, biological nitrogen fixation, endophytes, rhizobia, tripartite interactions

1. Introduction

The plant rhizosphere and phyllosphere is colonized by a wide range of epiphytic and endophytic microorganisms, and these microorganisms can establish beneficial, neutral, or detrimental associations of varying intimacy with their host plant. Recent developments in sequencing technologies have enabled us to study the composition and function of plant microbiomes, but these microbiomes are dynamic and differ among different plant tissues and in response to the environment. The microbiome can also be seen as "the second plant genome" and can consist of 10 times more genes than typical plant genomes [1]. Beneficial microorganisms that are associated with plants hold enormous potential to be developed into microbial fertilizers or microbial pesticides [2] and new biotechnological tools to improve the nutrient efficiency and stress tolerance of crops, and environmental sustainability of agroecosystems. Specific interactions between microbes and plants, such as the *rhizobium*-legume symbioses, are well understood, but the majority of the plant microbiome, and its contribution to the extended phenotype of the host, is not yet well defined.

Soybeans form interactions with nitrogen-fixing rhizobia, and this symbiosis plays a key role not only for the nitrogen (N) nutrition of the plant but also for agricultural productivity since soybean root residues provide N for other plants in crop rotations [3, 4]. Arbuscular mycorrhizal (AM) fungi colonize the root system of the majority of land plants, including soybeans; transfer nutrients such as phosphate (P), N, potassium (K), and other nutrients to their host plants; and improve the resistance of their host plants against abiotic (e.g., drought, salinity, and heavy metals) and biotic stresses [5]. In addition, soybeans are associated with bacterial or fungal endophytes that exhibit a wide range of plant growth promoting capabilities, including the production of phytohormones, an improved N nutrition through biological nitrogen fixation (diazotrophic endophytes), the biosynthesis of ACC (1-aminocyclopropane-1-carboxylate) deaminase, the capability to solubilize phosphate, and also the biosynthesis and release of antimicrobial metabolites or siderophores to inhibit the growth of pathogenic microorganisms [6].

The plant microbiome is a largely unexplored resource of beneficial microorganisms with diverse properties and a hidden potential to manipulate plant growth and success in stressful environments. However, while the symbiosis of soybeans with rhizobia and AM fungi is well characterized, the functional role of endophytes is only known for a limited number of isolates. Our functional understanding of these interactions is mainly based on experiments with individual symbionts, but there is increasing evidence that individual symbionts can also affect the interactions of the plant with other symbionts [7–10]. We summarize here the effects of different beneficial microbes on nutrient uptake, yield, and stress resistance of soybeans and identify knowledge gaps that hinder the application of these interactions to their full potential in soybean production systems.

2. Beneficial plant microbe interactions of soybean plants

2.1 Arbuscular mycorrhizal symbiosis

The arbuscular mycorrhizal (AM) symbiosis is arguably the most important symbiosis on earth and is formed by more than 65% of all known land plant species (n > 200,000), including all legumes and many other agronomically important crops, such as wheat, corn, and rice [11]. AM fungi are classified into the fungal subphylum Glomeromycota that consists of less than 350 fungal species [12]. AM fungi co-exist relatively morphologically unaltered with plants for more than 400 million years, and there is evidence that suggests that the AM symbiosis played a critical role for land plant evolution [13].

It is long known that AM fungi can increase the nutrient uptake of their host plant and are able to deliver substantial amounts of P, N, K, sulfur (S), and trace elements, such as copper (Cu) and zinc (Zn) to the plant. Many AM fungi also provide non-nutritional benefits for their host that are critical for plant survival or fitness and improve, for example, the resistance of plants against abiotic (e.g., drought, heavy metal, and salinity) and biotic (pathogens) stresses [5]. In return for these benefits, host plants transfer up to 20–25% of their photosynthetically

DOI: http://dx.doi.org/10.5772/intechopen.81396

derived carbohydrates to the fungal symbiont [14]. It was generally believed that carbon is transferred to the fungus in the form of hexoses [15], but recent evidence suggests that also fatty acids can move across the mycorrhizal interface to the fungal partner (**Figure 1**) [16–18].

AM fungi are ubiquitous in soils and can account for up to 50% of the microbial biomass in soils [19]. AM fungi form extensive hyphal networks in soils, and the extraradical mycelium (ERM) of the fungus acts as an extension of the root system and increases the nutrient absorbing surface of the root. The ERM with its mycorrhizosphere (interface between fungal hyphae and the soil) acts as an important conduit between microbial communities and the host plant [20] and can provide soil microbial communities with plant-derived carbon (C) inputs in large distance from the root. The mycorrhizosphere represents in soils an important ecological niche for diverse microbial communities that are specifically adapted to this mycorrhizosphere. According to estimates, the bacterial density in the mycorrhizosphere is 4–5 times higher than in the plant rhizosphere [21]. However, the presence of AM fungal mycelia does not only lead to quantitative but also to qualitative changes in the microbial community composition in soils [22]. The presence of AM fungal hyphae plays an important role in the bacterial community assembly during decomposition [22] and affects the access of members of these microbial communities to C sources during decomposition [23].

Within the host root, the fungus can not only spread intercellularly but also penetrate the root cortex intracellularly, and it forms here highly branched specialized structures called arbuscules that are separated from the plant symplast by the plant periarbuscular membrane [24]. Some AM fungal species also form vesicles, thick-walled, lipid-containing storage organs in the roots. Arbuscules are the site of nutrient exchange between the plant and the fungus, and both the fungal cell

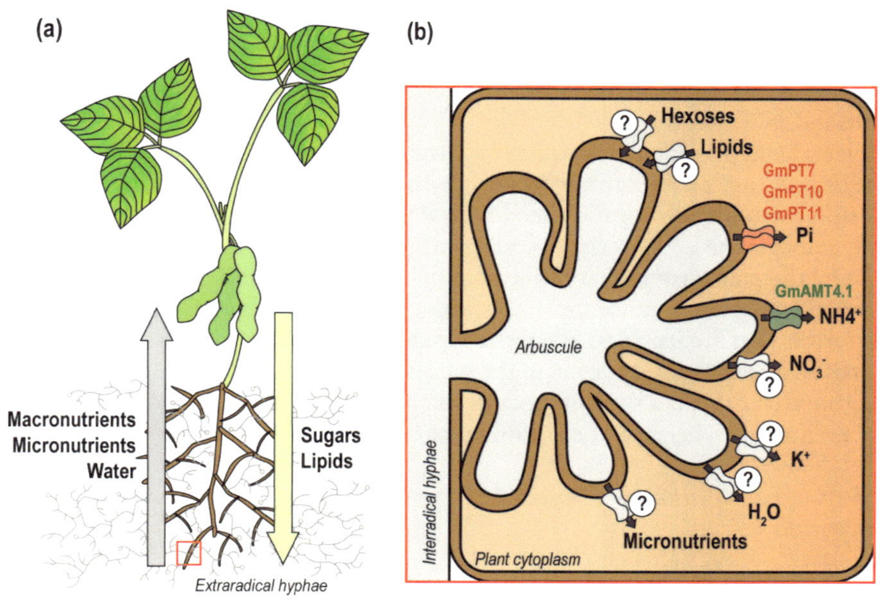

Figure 1.
Overview of the mycorrhizal nutrient uptake pathway in AM roots of soybean plants via the extraradical mycelium of the fungus (a) and the mycorrhizal interface consisting of the fungal arbuscule in root cortical cells surrounded by the periarbuscular membrane of the host (b). Both fungal cell membrane and plant periarbuscular membrane are characterized by the presence of mycorrhiza-specific transporters that play a critical role for the nutrient exchange across the mycorrhizal interface of soybean plants (e.g., GmPT7 or GmAMT4.1, see also below).

membrane and the plant periarbuscular membrane are characterized by the presence of specific transport proteins that play a critical role for the resource exchange between both partners (**Figure 1**) [15, 25, 26].

The colonization of host roots by AM fungi is based on a molecular dialog between both partners that facilitates partner recognition and triggers responses in both partners that are critical for the establishment of the symbiosis [27]. After fungal spore germination, an extensive hyphal branching in close proximity to host roots can be observed that is triggered by strigolactones and other compounds in root exudates [28]. After attachment to the host root surface and the differentiation of a fungal hyphopodium, the fungus penetrates the root, spreads with the help of a prepenetration apparatus [29], and forms arbuscules in the cells of the root cortex. This process is initiated by the release of lipochitooligosaccharides, or Myc factors, by the fungus that are perceived by specific receptors on the host root surface and trigger a cascade of molecular responses in the host root. The pathway is called the common symbiotic signaling pathway (CSSP), since similar responses can be observed after the perception of rhizobial Nod factors [27, 30]. A key role for the perception of fungal Myc or Nod factors by the rhizodermis plays the membrane-bound receptor-like kinase SYMRK that activates the mevalonate (MVA) biosynthetic enzyme HMGR1(3-hydroxy-3-methylglutaryl CoA reductase 1). A second set of CSSP proteins is located in the nuclear pore complex and includes the three nucleoporins NUP133, NUP85, and NENA; the ATP-powered Ca^{2+} pump MCA8; and cation channels encoded by CASTOR and POLLUX involved in the strong Ca^{2+} oscillations in the nucleus of rhizodermal cells that can be observed shortly after Myc factor perception. Another set of proteins is located in the nucleoplasm and decodes these Ca^{2+} signals [30, 31]. A Ca^{2+}/calmodulin-dependent protein kinase (CCaMK) phosphorylates with the help of calmodulin CYCLOPS, which then regulates gene expression either directly or through GRAS transcription factors such as NSP1, NSP2, and RAM1 [30–32]. The elucidation of the CSSP is mainly based on studies in the model legumes *Medicago truncatula* or *Lotus japonicus*, but the fact that the proteins of the CSSP are highly evolutionary conserved, and even present in plants that are unable to form AM interactions, suggests that this pathway is also established in soybeans.

Mycorrhizal plants have two pathways that are involved in the nutrient uptake from the soil: the "plant pathway" via high- and low-affinity transporters in root epidermis and root hairs or the "mycorrhizal pathway" that first involves the uptake of nutrients via the ERM of the fungus, transport to the arbuscules, and then the uptake by the plant from the interfacial apoplast through specialized transporters in the periarbuscular membrane. In response to the colonization with AM fungi, transporters that are involved in the plant pathway are often down-regulated, while mycorrhiza-specific transporters in the periarbuscular membrane are induced [33], indicating that there is a shift in the nutrient acquisition strategy and that the mycorrhizal pathway can become the dominant pathway for nutrient uptake [34, 35].

2.1.1 *Importance of arbuscular mycorrhizal fungi for yield and nutrient uptake of soybeans*

Under both greenhouse and field conditions, increases in nutrient content, yield, and overall fitness of soybeans in response to an AM colonization can be observed [36, 37], and soybean yields are significantly correlated to the colonization of the roots with AM fungi [38]. Many reports clearly demonstrate the positive effects of AM fungi on the nutrient uptake of soybeans and here particularly on the uptake of phosphorus (P) and nitrogen (N) [39–41]. However, the effects can differ greatly among AM fungi. Our own studies demonstrated, for example, that while the AM fungus *Rhizophagus irregularis* can increase the P nutrition of soybeans with low or high P acquisition efficiency, *Glomus custos* had no effect and *Glomus aggregatum* even led to slight growth depressions under medium P supply conditions [39].

Some of the observed differences among these AM fungi seem to be related to the impact of the AM fungus on plant P transporter expression. Fourteen genes of the Pht1 family have been identified in soybeans [42], and three of these transporters show high expression levels in AM roots [43]. While the colonization of the roots with *R. irregularis* led to the downregulation of *GmPt4*, a high-affinity P uptake transporter that is presumably involved in the uptake of P from the soil via the plant uptake pathway was the expression of *GmPt9*, and *GmPt10* upregulated in AM roots. *GmPt9* and *GmPt10* cluster with the mycorrhiza-inducible P transporters *OsPt11* of *Oryza sativa* (rice) and *MtPt4* of *Medicago truncatula* that play a critical role for the P uptake from the mycorrhizal interface [26, 44]. *GmPt9* was upregulated by *G. aggregatum* and *R. irregularis*, but *GmPt10* was only upregulated by *R. irregularis*, indicating that this transporter is involved in the P uptake from the interface and that *GmPt10* expression can serve as an indicator for mycorrhizal P benefits in soybean plants. *GmPt7*, another soybean P transporter, shows a high expression in cells with mature and active arbuscules but is not expressed in cells with collapsed and degenerated arbuscules, suggesting that this transporter may also play a role for the P transport across the AM interface. However, *GmPt7* is not a mycorrhiza-specific transporter and is also expressed in columella cells of root caps and in lateral root primordia of nonmycorrhizal roots [45]. Similarly, out of the 16 ammonium (NH_4^+) transporters of soybean, five transporters are mycorrhiza-inducible, and one of them, *GmAMT4.1*, is specifically expressed in arbusculated cells (**Figure 1**), indicating that this transporter could be involved in the NH_4^+ transport across the AM interface [46]. There is evidence from the model legume *Medicago truncatula* that AM fungi can also improve the acquisition of other macronutrients such as potassium (K) or sulfur [47, 48]. K deficiency is a common problem in soybeans and can lead to yellowing of the leaves, stunted growth, and reduced yields and can become particularly severe under drought stress. Although transcriptional and physiological responses to K deprivation have been studied in other legumes [49], whether AM fungi also play a role in the K acquisition of soybean plants is not yet known.

2.1.2 Importance of arbuscular mycorrhizal fungi for the stress resistance of soybeans

AM fungi can also increase the resistance of soybeans against other abiotic stresses such as drought, salinity, or soil contaminations. It is known for several decades that the AM colonization can improve the tolerance of soybeans against drought [50]. AM fungi can influence leaf water potential, solute accumulation, and oxidative stress of soybeans under drought stress [51] and delay nodule senescence triggered by water deprivation [52]. In mycorrhizal soybeans, plasma membrane aquaporins were down-regulated in response to drought stress, and this could reduce the permeability of membranes for water and contribute to water conservation [53]. In addition, both fungal and plant mitogen-activated protein kinases (MAPKs) are upregulated in AM soybean plants under drought stress. MAPK cascades are known to regulate many cellular processes in response to various stimuli, including abiotic and biotic stresses [54]. AM fungi also improve the tolerance of soybeans against salinity. AM plants had a higher biomass and proline concentrations in roots, but reduced proline and Na concentrations in the shoot under salt stress. When the fungus was pretreated with NaCl, the alleviating effects were even stronger, indicating that the acclimation of the fungus to salinity may play a role for the stress response [55]. AM fungi can also improve the tolerance of soybeans against arsenic [56] and aluminum [57] by reducing the uptake of these toxic metals.

Soybean yield and productivity is also threatened by many fungal or bacterial diseases and soil inhabiting nematodes. Soybean cyst nematodes (SCNs, *Heterodera glycines*), brown spot (*Septoria glycines*), charcoal rot (*Macrophomina phaseolina*), rot and stem rot (*Phytophthora sojae*), and soybean rust (*Phakopsora pachyrhizi* and *P. meibomiae)* are among the most important pathogens of soybeans and cause substantial yield losses in the U.S. [58]. SCNs are often responsible for hidden yield losses, since soil infestations remain often undetected until they become severe. SCN can spread easily from field to field via soil movements with machinery, wind, or by humans and can now be detected in 90% of the soybean producing states in the U.S. [59]. SCN infestations can lead to yield losses of more than 30% and are responsible for about $ 1.5 billion in soybean crop damage each year in the U.S. AM fungi can protect soybeans against a wide range of pathogens, including fungi, bacteria, nematodes, or insects [60] and reduce the SCN egg population in soils by 70% [61]. The positive impact of AM fungi on biotic stresses has been attributed to the overall positive effect on nutrient uptake and a damage compensation effect, the competition for root space and soil nutrients, induced systemic resistance (ISR), and altered rhizosphere interactions. In addition, AM fungi form extensive hyphal networks in soils and can connect plants of the same or of different plant species by common mycelial networks (CMNs). CMNs play an important role in the plant-to-plant communication and can transfer infochemicals and warning signals from infested plants to uninfested plants and stimulate defense reactions in these plants [62].

2.2 Nitrogen-fixing symbiosis with rhizobia

Most legume plants are able to interact with N-fixing bacteria called rhizobia that are able to reduce atmospheric dinitrogen (N_2) into ammonia (NH_3) in specialized root nodules. The symbiosis evolved in legumes between 25 and 50 million years ago [63, 64] and plays an important role for plant nitrogen (N) nutrition. Rhizobia can contribute up to 70% to the total N nutrition, grain legumes can gain up to 300 kg N, and legume trees (e.g., *Acacia* sp.) up to 600 kg N per ha and year from these interactions [4, 65]. Free living rhizobia produce Nod factors that are perceived by plant roots and act as triggers for the common symbiotic signaling pathway (CSSP, see above). Nod factors are also lipochitooligosaccharides that are composed of chitin chains with various lipid modifications. Chitin is the main constituent of fungal but not of bacterial cell walls, and the functional and structural similarities between Nod and Myc factors have led to the assumption that rhizobia adopted the evolutionary far more ancient (~ 450 million years) CSSP to establish this endosymbiotic interaction with legumes [66]. Nod factors stimulate the curling of root hairs, and entrapped bacteria within these curls are transported within infection threads to the inner zone of developing root nodules. Inside of cortical cells, the rhizobia divide and multiply, and are released into vesicles, called symbiosomes, in which they differentiate to fully functional bacteroids. One or more differentiated bacteroids are surrounded by the plant symbiosome membrane, which represents a barrier by which the host plant can control the movement of solutes to the bacteroids through specialized transporters or channels [67].

Bacteroids express the nitrogenase complex that consists of six protein subunits (two each of NifH, NifD, and NifK), two [4Fe–4S], two (Fe_8S_7) iron-sulfur clusters, and two iron-molybdenum cofactors (Fe_7MoS_9N) called FeMoco, which catalyze the N_2 reduction to NH_3 [68]. The nitrogenase metallocenters are all oxygen-labile and must operate in an environment with a low level of free oxygen, and nodules provide their bacterial symbionts with this oxygen-reduced environment for optimum N fixation [69]. N fixation by bacteroids is a highly energy consuming process, and rapid respiration in the bacteroids is necessary to produce the 16 ATP required for the conversion of each atmospheric N_2 into two NH_3.

$$N_2 + 8H^+ + 8e^- + 16ATP = 2NH_3 + H_2 + 16ADP + 16P_i$$

The product of biological N fixation (BNF) is ammonia, which diffuses out of the bacteroids into the acidic symbiosome space and is here protonated to ammonium. The symbiosome membrane is energized by an H^+-ATPase, which pumps protons into the symbiosome space and thereby promotes the uptake of NH_3/NH_4^+ into the plant cytosol, where NH_4^+ is rapidly assimilated into amino acids, and the ureides, allantoin and allantoic acid [69]. A candidate for the uptake of NH_4^+ from the symbiosome space is *NOD26*, which was first identified in soybeans [70]. NOD26 belongs to the major intrinsic protein/aquaporin (MIP/AQP) channel family and is exclusively localized in the symbiosome membrane [67]. The ureides, allantoin and allantoic acid, serve as the dominant long-distance transport molecule for N from the root nodules to the shoots [71, 72]. Cortex cells and the vascular endodermis of nodules express *GmUPS1*—1 and *GmUPS1–2*, which play a role for the transport of allantoin and allantoic acid out of the root nodules to the sink organs. RNA*i* knockouts of these proteins accumulate ureides in the root nodules and show a reduced N transport to the shoots [73].

BNF is an energy expensive process, which requires 16 ATP to fuel the reduction of one N_2. Plants allocate up to 30% of their photosynthetically fixed C to rhizobia [74], which is oxidized in the bacteroids to ATP. The N_2 fixation rate of rhizobia is higher when the nodules receive more C, suggesting that the allocation of C to nodules is a limiting factor for BNF. Transgenic *Medicago sativa* plants that overexpress a sucrose phosphate synthase, a key enzyme for sucrose biosynthesis in plants, show higher C contents in nodules, more and larger nodules per plant, and an enhanced nitrogenase activity of the root nodules [75]. Free-living rhizobia can grow on a variety of different sugars, including mono- and disaccharides, but the absence of transporters for these sugars in bacteroids suggests that rhizobia in symbiosis take up dicarboxylates and here particularly malate from the symbiosome space. The C4-dicarboxylate transport system that is localized in the inner bacteroid membrane is encoded by the *dctA* gene, has a high mobility for malate, and is essential for symbiotic nitrogen fixation [76]. Although the mechanisms of N fixation and assimilation are well documented, key steps are still unknown. For example, little is known about the C metabolism inside nodules, the regulatory steps that control the C export to rhizobia, and the proteins involved in the C and N transport between partners. Recent evidence in the model legumes *M. truncatula* and *Lotus japonicus* suggests that sucrose transporters from the Sugar Will Eventually be Exported Transporter (SWEET) family could be involved in the sucrose efflux from the phloem toward nodulated cells [77].

2.2.1 Significance of rhizobia for soybean agriculture

According to estimates, soybeans with their rhizobia populations fix around 20 million tons of N each year, and this has an enormous influence on agricultural productivity, not only on soybeans, but also on other crops in crop rotation systems [3, 4]. Soybean residues in the soil enrich the soil with N, improve soil organic matter, and can lead to yield increases in non-legume crops that follow soybeans. Crop rotations or intercropping systems of cereals with legumes can result in higher crop yields without fertilizer additions [78]. However, conventional agricultural management practices and other anthropogenic factors can have a negative impact on rhizobial function. In addition, excessive tillage, applications of higher N fertilizer dosages, and extended fallow periods can also have detrimental effects on rhizobia populations in soils. As a consequence, integrating this symbiosis more efficiently in modern agricultural practices is crucial to limit the amount of fertilizers used and make agriculture more environmentally sustainable. Exploring ecologically best fitted ecoregions for soybeans and best adapted soybean cultivars will help farmers to produce more yield with reduced inputs. Rhizobial strains differ in their efficacy in symbiosis with different soybean cultivars, and the input of N into agricultural systems can be increased by the inoculation of legumes with optimized rhizobia for different environments [65]. The development of better inoculation strategies and specifically adapted rhizobia for different soybean cultivars could reduce the dependency of farmers on agrochemicals and enhance food security [65].

2.3 Tripartite symbiosis with arbuscular mycorrhizal fungi and rhizobia

In natural environments, legume roots form tripartite interactions and are simultaneously colonized by both AM fungi and rhizobia [7, 79]. Tripartite interactions have been shown to improve plant productivity, seed yield, P and N acquisition, and photosynthetic rates [10, 80, 81]. The rhizobial nitrogenase complex requires at least 16 ATP to reduce one N_2 molecule into two NH_3. Consequently, nodules act as strong P sinks in legume root systems to provide sufficient P resources to the bacteroids for optimum BNF [79, 82]. Since AM fungi are able to improve the P nutrition of legume plants, AM fungi can increase the BNF by root nodules by at least 50% [10]. Nonmycorrhizal soybean plants have lower nodule numbers and weights and particularly under low P supply lower N fixation rates [7, 83]. AM fungi can also provide their hosts with microelements that are essential for N_2 fixation, including zinc, iron, manganese, and molybdenum [84, 85].

AM fungi and rhizobial bacteria can act synergistically and can improve plant productivity, seed yield, and grain quality [7, 10, 81]. However, the prior inoculation by either rhizobia or AM fungi can also reduce the subsequent colonization by the other symbiont [86]. Plants control the extent of root colonization by both symbionts by an autoregulatory mechanism, possibly to limit the high C costs associated with these interactions [83, 87]. Whether AM fungi and rhizobia interact antagonistically or synergistically depends on the environmental context [81] and the compatibility between symbiotic partners [10, 88]. For example, the rhizobial strain STM 7183 is more compatible with the AM fungus *Rhizophagus clarus* and leads to higher nodulation rates, nitrogenase activities, and plant growth responses than STM 7282 [10]. Similarly, plant productivity and seed yields of nodulated soybeans were higher when the plants were co-inoculated with the AM fungus *Rhizophagus irregularis* than with *Acaulospora tuberculata* or *Gigaspora gigantea* [88]. Soybean cultivars also differ in their ability to benefit from their microbial communities [89]. Consequently, the symbiotic efficiency should be integrated into soybean breeding programs, and AM fungi and N-fixing bacteria with high compatibility should be identified to improve the productivity and stress resistance of soybeans and other legumes.

Both interactions are costly, and the host plant allocates up to 20% of its photosynthetically fixed C to its fungal [14, 90] and up to 30% to its N-fixing symbionts (**Figure 2**) [74]. C acts as an important trigger for symbiotic functioning, and a reduction in the C fluxes to the symbionts decreases BNF by rhizobia [91], and P and N uptake and transport by AM fungi [92–94]. Considering the high C costs of these symbioses for the host, plants are under a selective pressure to strongly regulate the C fluxes to both root symbionts, but these control mechanisms are currently poorly understood. Resource exchange between host and AM fungi is controlled by a reciprocal reward mechanism that is driven by biological market dynamics [95].

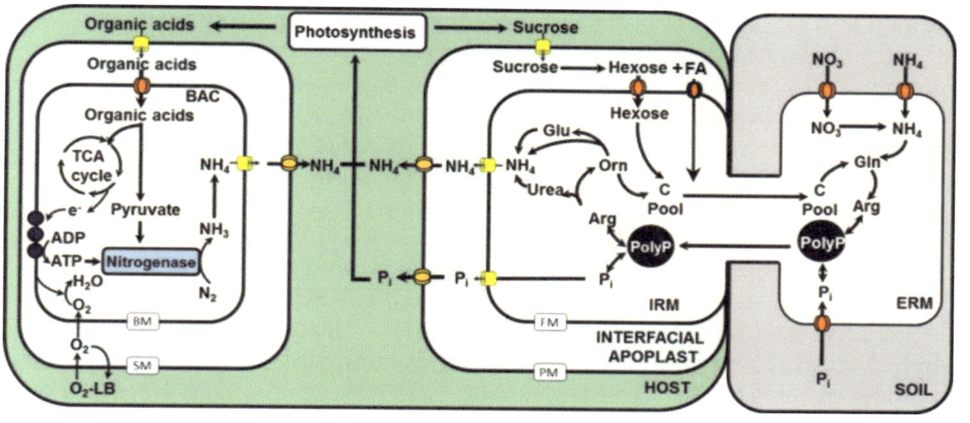

Figure 2.
Transport and nutrient exchange pathways in the symbiosis with N-fixing bacteria (BAC) and AM fungi (IRM and ERM). Abbreviations: BAC, N-fixing bacteroid; BM, Bacteroid membrane; ERM, Extraradical mycelium; FA, Fatty acids; FM, Fungal plasma membrane; IRM, Intraradical mycelium; PM, Periarbuscular membrane; SM, Symbiosome membrane.

Our own results recently demonstrated that similar mechanisms may also control the resource to C exchange in tripartite interactions; *Medicago* plants allocate C to the different root symbionts in tripartite interactions in response to nutrient demand conditions; and the AM fungus becomes a stronger competitor for C resources from the host, when the fungal partner has access to N [79].

AM fungi have stronger effects on plant gene expression than rhizobia [96], but our current understanding of the molecular mechanisms involved in the C allocation to individual root symbionts is limiting. An overexpression of a leaf sucrose phosphate synthase of *M. sativa* increases starch production, allowing the plant to allocate more photosynthates to root nodules and consequently improved nitrogenase activity and overall plant growth [75]. There is evidence that suggests that sucrose transporters (SUTs) could be involved in the regulation of beneficial C fluxes toward the fungal symbiont [97], and the expression of *MtSUT2* and *MtSUT4–1* has been shown to be positively correlated to the C allocation to different symbiotic partners in tripartite interactions [79]. *MtSWEET1b* and *MtSWEET6* of the Sugars Will Eventually be Exported Transporter family (SWEET) are highly expressed in AM roots, and preferentially transport hexoses such as glucose, and could be involved in the transport of hexoses or fatty acids across the mycorrhizal interface to the fungal partner [79, 98]. *MtSWEET11* is specifically expressed in root nodules and could be involved in the sugar distribution within root nodules, but loss-of-function mutants indicate that *MtSWEET11* is not essential for BNF [99]. A better understanding of these processes is critical because it may be key to improve the resource exchange between plants and symbionts and ultimately to enhance productivity of agronomically important legumes.

2.4 Symbiosis with endophytic bacteria or fungi

Endophytes are defined as organisms that live inside plant hosts for at least part of their lives, without causing apparent disease symptoms in the host as a result of this colonization [100]. Fungal and bacterial endophytes are nearly ubiquitous across all groups of vascular plants [101], but there is a large biological diversity among endophytes, and it is not rare for some plant species to host hundreds of different endophytic species [102]. Fungal endophytes have been shown to enhance growth and seed production or protect against environmental stresses such as drought or P deficiency or provide defense against herbivory through the synthesis of various biologically active metabolites, such as alkaloids. In soybeans diverse communities of fungal endophytes can be found, and several of these endophytes have plant growth-promoting capabilities and enhance, for example, soybean growth in nickel- or copper-contaminated soils by reducing the levels of stress-related phytohormones such as abscisic acid and jasmonic acid [103]; or increase glutathione activities and thereby reduce oxidative stress [104]. The inoculation of soybean plants with fungal endophytes can also lead to higher shoot biomasses, chlorophyll contents, and photosynthetic rates compared to noninoculated soybeans under salt stress and decrease the abundances of SCN in soils [105].

Soybeans host also a diverse group of bacterial endophytes, and many endophytic bacteria have plant growth-promoting capabilities [106], such as the ability to produce plant growth hormones, or ACC (1-aminocyclopropane-1-carboxylate) deaminase, solubilize phosphate, or release antimicrobial metabolites or siderophores that can inhibit the growth of pathogenic microorganisms. ACC deaminase reduces the levels of ethylene, an important stress hormone in plants. Several endophytic bacteria are also diazotrophs and have like rhizobia bacteria the ability to fix N. Bacterial endophytes also interact with rhizobia bacteria and can enhance root nodulation and activity, and as a consequence, the N content of soybean plants [107]. The dual inoculation with rhizobia and a salt-tolerant bacterial endophyte led to synergistic responses and promoted the fitness of soybean plants under salt stress [108].

3. Important research gaps and future challenges

Beneficial plant microbe interactions with AM fungi, rhizobia, or bacterial and fungal endophytes have enormous potential to improve plant growth and nutrient uptake in stressful environments and to increase the environmental sustainability of soybean agriculture. However, while the beneficial effects of AM fungi and rhizobia on soybean productivity are long known, the effect of only a small number of endophytes is currently known. The plant microbiome is a still unexplored resource of microorganisms with a so far hidden potential to promote plant growth and success under abiotic or biotic stress conditions, and with unknown effects on the plant phenotype.

The obligate lifestyle of AM fungi has made for a long time the production of fungal inoculum in large quantities difficult, but the development of sterile transgenic root organ cultures has led to an increased commercialization of AM fungal inocula for the utilization in agroecosystems [109]. Although increases in yield and biomass have been reported in different crops after inoculation with these inocula [36, 110], in other studies, inconsistent or neutral effects were observed [111]. AM fungi differ in the benefit that they provide for their host plant [112], and mycorrhizal growth responses are highly context dependent. Several factors can alter the success of AM fungal inoculation in agroecosystems, including plant/fungal compatibility, the degree of competition with the native microbial population, or timing of inoculation [113]. All these aspects need to be taken into consideration to find the most adapted and specific conditions for an efficient use of AM fungal inocula in a given field or for a certain crop. Our current understanding of the effect of beneficial plant microbes on soybeans is mainly based on studies with single symbiont, but plant productivity and stress resistance in agroecosystems depend on diverse microbial communities and the interactions among the different microorganisms in these communities. Identifying and characterizing the molecular mechanisms responsible for the functioning of different plant microbe interactions is crucial to harness these symbiotic microorganisms in agroecosystems. Currently, most knowledge is gathered on model legumes, such as *Medicago truncatula*, but the information about soybeans is limited. However, the accumulation of genomic and transcriptomic data, along with the development of molecular tools such as stable transformations, e.g., [114], CRISPR-Cas9 system [115], or mutant populations, will provide us with a better understanding of these interactions in soybeans.

Acknowledgements

We wish to acknowledge funding from the USDA (2017-67014-26530), the SD Soybean Research and Promotion Council, the Agricultural Experiment Station at SDSU, and the funding from the North Carolina Agricultural Research Service (NCARS) for K.G.

Conflict of interest

The authors have no conflict of interest.

Author details

Arjun Kafle[1], Kevin Garcia[1,2], Vincent Peta[1], Jaya Yakha[1], Alex Soupir[1] and Heike Bücking[1*]

1 Biology and Microbiology Department, McFadden Biostress, South Dakota State University, Brookings, USA

2 Department of Crop and Soil Sciences, Center for Integrated Fungal Research, North Carolina State University, Raleigh, USA

*Address all correspondence to: heike.bucking@sdstate.edu

References

[1] Turner TR, James EK, Poole PS. The plant microbiome. Genome Biology. 2013;**14**:209. DOI: 10.1186/gb-2013-14-6-209

[2] Vryzas Z. The plant as metaorganism and research on next-generation systemic pesticides – Prospects and challenges. Frontiers in Microbiology. 2016;7. DOI: 10.3389/fmicb.2016.01968

[3] Peoples MB, Brockwell J, Herridge DF, Rochester IJ, Alves BJR, Urquiaga S, et al. The contributions of nitrogen-fixing crop legumes to the productivity of agricultural systems. Symbiosis. 2009;**48**:1-17. DOI: 10.1007/bf03179980

[4] Herridge DF, Peoples MB, Boddey RM. Global inputs of biological nitrogen fixation in agricultural systems. Plant and Soil. 2008;**311**:1-18. DOI: 10.1007/s11104-008-9668-3

[5] Smith SE, Read DJ. Mycorrhizal symbiosis. 3rd ed. New York: Academic Press; 2008. DOI: 10.1016/B978-0-12-370526-6.X5001-6

[6] Santoyo G, Moreno-Hagelsieb G, Del Carmen Orozco-Mosqueda M, Glick BR. Plant growth-promoting endophytes. Microbiological Research. 2016;**183**:92-99. DOI: 10.1016/j.micres.2015.11.008

[7] Püschel D, Janouskova M, Voriskova A, Gryndlerova H, Vosatka M, Jansa J. Arbuscular mycorrhiza stimulates biological nitrogen fixation in two *Medicago* spp. through improved phosphorus acquisition. Frontiers in Plant Science. 2017;**8**. DOI: 10.3389/fpls.2017.00390

[8] Ossler JN, Zielinski CA, Heath KD. Tripartite mutualism: Facilitation or trade-offs between rhizobial and mycorrhizal symbionts of legume hosts. American Journal of Botany. 2015;**102**:1332-1341. DOI: 10.3732/ajb.1500007

[9] Bulgarelli RG, Correia Marcos FC, Ribeiro RV, Lopez De Andrade SA. Mycorrhizae enhance nitrogen fixation and photosynthesis in phosphorus starved soybean (*Glycine max* L. Merrill). Environmental and Experimental Botany. 2017;**140**:26-33. DOI: 10.1016/j.envexpbot.2017.05.015

[10] Bournaud C, James EK, De Faria SM, Lebrun M, Melkonian R, Duponnois R, et al. Interdependency of efficient nodulation and arbuscular mycorrhization in *Piptadenia gonoacantha*, a Brazilian legume tree. Plant, Cell & Environment. 2017. DOI: 10.1111/pce.13095

[11] Wang B, Qiu Y-L. Phylogenetic distribution and evolution of mycorrhizae in land plants. Mycorrhiza. 2006;**16**:299-363. DOI: 10.1007/s00572-005-0033-6

[12] Öpik M, Zobel M, Cantero JJ, Davison J, Facelli JM, Hiiesalu I, et al. Global sampling of plant roots expands the described molecular diversity of arbuscular mycorrhizal fungi. Mycorrhiza. 2013;**23**:411-430. DOI: 10.1007/s00572-013-0482-2

[13] Delaux P-M, Radhakrishnan GV, Jayaraman D, Cheem J, Malbreil M, Volkening JD, et al. Algal ancestor of land plants was preadapted for symbiosis. Proceedings of the National Academy of Sciences of the United States of America. 2015;**112**:13390-13395. DOI: 10.1073/pnas.1515426112

[14] Wright DP, Read DJ, Scholes JD. Mycorrhizal sink strength influences whole plant carbon balance of *Trifolium repens* L. Plant, Cell and Environment. 1998;**21**:881-891. DOI: 10.1046/j.1365-3040.1998.00351.x

[15] Helber N, Wippel K, Sauer N, Schaarschmidt S, Hause B, Requena N. A versatile monosaccharide transporter that operates in the arbuscular mycorrhizal fungus *Glomus* sp is crucial for the symbiotic relationship with plants. The Plant Cell. 2011;**23**:3812-3823. DOI: 10.1105/tpc.111.089813

[16] Bravo A, Brands M, Wewer V, Dörmann P, Harrison MJ. Arbuscular mycorrhiza-specific enzymes FatM and RAM2 fine-tune lipid biosynthesis to promote development of arbuscular mycorrhiza. New Phytologist. 2017;**214**:1631-1645. DOI: 10.1111/nph.14533

[17] Keymer A, Pimprikar P, Wewer V, Huber C, Brands M, Bucerius SL, et al. Lipid transfer from plants to arbuscular mycorrhiza fungi. eLife. 2017;**6**. DOI: 10.7554/eLife.29107

[18] Luginbuehl LH, Menard GN, Kurup S, Van Erp H, Radhakrishnan GV, Breakspear A, et al. Fatty acids in arbuscular mycorrhizal fungi are synthesized by the host plant. Science. 2017;**356**:1175-1178. DOI: 10.1126/science.aan0081

[19] Ryan MH, Graham JH. Is there a role for arbuscular mycorrhizal fungi in production agriculture? Plant and Soil. 2002;**244**:263-271. DOI: 10.1023/a:1020207631893

[20] Kaiser C, Kilburn MR, Clode PL, Fuchslueger L, Koranda M, Cliff JB, et al. Exploring the transfer of recent plant photosynthates to soil microbes: Mycorrhizal pathway vs direct root exudation. New Phytologist. 2014;**205**:1537-1551. DOI: 10.1111/nph.13138

[21] Heinonsalo J, Jörgensen KS, Haahtela K, Sen R. Effects of *Pinus sylvestris* root growth and mycorrhizosphere development on bacterial carbon source utilization and hydrocarbon oxidation in forest and petroleum-contaminated soils. Canadian Journal of Microbiology. 2000;**46**:451-464. DOI: 10.1139/cjm-46-5-451

[22] Nuccio EE, Hodge A, Pett-Ridge J, Herman DJ, Weber PK, Firestone MK. An arbuscular mycorrhizal fungus significantly modifies the soil bacterial community and nitrogen cycling during litter decomposition. Environmental Microbiology. 2013;**15**:1870-1881. DOI: 10.1111/1462-2920.12081

[23] Herman DJ, Firestone MK, Nuccio E, Hodge A. Interactions between an arbuscular mycorrhizal fungus and a soil microbial community mediating litter decomposition. FEMS Microbiology Ecology. 2012;**80**:236-247. DOI: 10.1111/j.1574-6941.2011.01292.x

[24] Brundrett M. Diversity and classification of mycorrhizal associations. Biological Reviews. 2004;**79**:473-495. DOI: 10.1017/s1464793103006316

[25] Garcia K, Doidy J, Zimmermann SD, Wipf D, Courty P-E. Take a trip through the plant and fungal transportome of mycorrhiza. Trends in Plant Science. 2016;**21**:937-950. DOI: 10.1016/j.tplants.2016.07.010

[26] Javot H, Penmetsa RV, Terzaghi N, Cook DR, Harrison MJ. A *Medicago truncatula* phosphate transporter indispensable for the arbuscular mycorrhizal symbiosis. Proceedings of the National Academy of Sciences, USA. 2007;**104**:1720-1725. DOI: 10.1073/pnas.0608136104

[27] Delaux P-M, Séjalon-Delmas N, Bécard G, Ané J-M. Evolution of the plant–microbe symbiotic 'toolkit'. Trends in Plant Science. 2013;**18**:298-304. DOI: 10.1016/j.tplants.2013.01.008

[28] Akiyama K, Hayashi H. Strigolactones: Chemical signals for fungal symbionts and parasitic weeds in plant roots. Annals of Botany. 2006;**97**:925-931. DOI: 10.1093/aobmc1063

[29] Genre A, Chabaud M, Timmers T, Bonfante P, Barker DG. Arbuscular mycorrhizal fungi elicit a novel intracellular apparatus in *Medicago truncatula* root epidermal cells before infection. The Plant Cell. 2005;**17**: 3489-3499. DOI: 10.1105/tpc.105.035410

[30] Parniske M. Arbuscular mycorrhiza: The mother of plant root endosymbioses. Nature Reviews Microbiology. 2008;**6**:763-775. DOI: 10.1038/nrmicro1987

[31] Hohnjec N, Czaja-Hasse LF, Hogekamp C, Küster H. Pre-announcement of symbiotic guests: Transcriptional reprogramming by mycorrhizal lipochitooligosaccharides shows a strict co-dependency on the GRAS transcription factors NSP1 and RAM1. BMC Genomics. 2015;**16**:994. DOI: 10.1186/s12864-015-2224-7

[32] Smit P, Raedts J, Portyanko V, Debelle F, Gough C, Bisseling T, et al. NSP1 of the GRAS protein family is essential for rhizobial nod factor-induced transcription. Science. 2005;**308**:1789-1791. DOI: 10.1126/science.1111025

[33] Grunwald U, Guo WB, Fischer K, Isayenkov S, Ludwig-Müller J, Hause B, et al. Overlapping expression patterns and differential transcript levels of phosphate transporter genes in arbuscular mycorrhizal, P_i-fertilised and phytohormone-treated *Medicago truncatula* roots. Planta. 2009;**229**: 1023-1034. DOI: 10.1007/s00425-008-0877-z

[34] Li H, Smith FA, Dickson S, Holloway RE, Smith SE. Plant growth depressions in arbuscular mycorrhizal symbioses: Not just caused by carbon drain? New Phytologist. 2008;**178**:852-862. DOI: 10.1111/j.1469-8137.2008.02410.x

[35] Smith SE, Jakobsen I, Gronlund M, Smith FA. Roles of arbuscular mycorrhizas in plant phosphorus nutrition: Interactions between pathways of phosphorus uptake in arbuscular mycorrhizal roots have important implications for understanding and manipulating plant phosphorus acquisition. Plant Physiology. 2011;**156**:1050-1057. DOI: 10.1104/pp.111.174581

[36] Cely MVT, De Oliveira AG, De Freitas VF, De Luca MB, Barazetti AR, Dos Santos IMO, et al. Inoculant of arbuscular mycorrhizal fungi (*Rhizophagus clarus*) increase yield of soybean and cotton under field conditions. Frontiers in Microbiology. 2016;**7**:720. DOI: 10.3389/fmicb.2016.00720

[37] Afolayan ET, Eguavon MI. Morphological and yield assessment of soybean (*Glycine max* L.) as influenced by arbuscular mycorrhizal fungi and other soil amendments. Journal of Agricultural Science and Food Research. 2017;**8**:193

[38] Isobe K, Higo M, Kondo T, Sato N, Takeyama S, Torigoe Y. Effect of winter crop species on arbuscular mycorrhizal fungal colonization and subsequent soybean yields. Plant Production Science. 2014;**17**:260-267. DOI: 10.1626/pps.17.260

[39] Wang X, Zhao S, Bücking H. Arbuscular mycorrhizal growth responses are fungal specific but do not differ between soybean genotypes with different phosphate efficiency. Annals of Botany. 2016;**18**:11-21. DOI: 10.1093/aob/mcw074

[40] Wang X, Yan X, Liao H. Genetic improvement for phosphorus efficiency in soybean: A radical approach. Annals

of Botany. 2010;**106**:215-222. DOI: 10.1093/aob/mcq029

[41] Liu H, Song F, Liu S, Li X, Liu F, Zhu X. Arbuscular mycorrhiza improves nitrogen use efficiency in soybean grown under partial root-zone drying irrigation. Archives of Agronomy and Soil Science. 2018. DOI: 10.1080/03650340.2018.1493724

[42] Qin L, Guo YX, Chen LY, Liang RK, Gu MA, Xu GH, et al. Functional characterization of 14 Pht1 family genes in yeast and their expressions in response to nutrient starvation in soybean. PLoS One. 2012;7:e47726. DOI: 10.1371/journal.pone.0047726

[43] Tamura Y, Kobae Y, Mizuno T, Hata S. Identification and expression analysis of arbuscular mycorrhiza-inducible phosphate transporter genes of soybean. Bioscience, Biotechnology and Biochemistry. 2012;**76**:309-313. DOI: 10.1271/bbb.110684

[44] Kobae Y, Hata S. Dynamics of periarbuscular membranes visualized with a fluorescent phosphate transporter in arbuscular mycorrhizal roots of rice. Plant and Cell Physiology. 2010;**51**:341-353. DOI: 10.1093/pcp/pcq013

[45] Inoue Y, Kobae Y, Omoto E, Tanaka A, Banba M, Takai S, et al. The soybean mycorrhiza-inducible phosphate transporter gene, *GmPT7*, also shows localized expression at the tips of vein endings of senescent leaves. Plant and Cell Physiology. 2014;**55**:2102-2111. DOI: 10.1093/pcp/pcu138

[46] Kobae Y, Tamura Y, Takai S, Banba M, Hata S. Localized expression of arbuscular mycorrhiza-inducible ammonium transporters in soybean. Plant and Cell Physiology. 2010;**51**: 1411-1415. DOI: 10.1093/pcp/pcq099

[47] Sieh D, Watanabe M, Devers EA, Brueckner F, Hoefgen R, Krajinski F. The arbuscular mycorrhizal symbiosis influences sulfur starvation responses of *Medicago truncatula*. New Phytologist. 2013;**197**:606-616. DOI: 10.1111/nph.12034

[48] Garcia K, Chasman D, Roy S, Ané J-M. Physiological responses and gene co-expression network of mycorrhizal roots under K deprivation. Plant Physiology. 2017;**173**:1811-1823. DOI: 10.1104/99.16.01959

[49] Garcia K, Ané J-M. Polymorphic responses of *Medicago truncatula* accessions to potassium deprivation. Plant Signaling & Behavior. 2017;**12**:e1307494. DOI: 10.1080/15592324.2017.1307494

[50] Busse M, Ellis J. Vesicular–arbuscular mycorrhizal (*Glomus fasciculatum*) influence on soybean drought tolerance in high phosphorus soil. Canadian Journal of Botany. 1985;**63**:2290-2294. DOI: 10.1139/b85-327

[51] Porcel R, Ruiz-Lozano JM. Arbuscular mycorrhizal influence on leaf water potential, solute accumulation, and oxidative stress in soybean plants subjected to drought stress. Journal of Experimental Botany. 2004;**55**:1743-1750. DOI: 10.1093/jxb/erh188

[52] Ruiz-Lozano JM, Collados C, Barea JM, Azcón R. Arbuscular mycorrhizal symbiosis can alleviate drought-induced nodule senescence in soybean plants. New Phytologist. 2001;**151**:493-502

[53] Porcel R, Aroca R, Azcon R, Ruiz-Lozano JM. PIP aquaporin gene expression in arbuscular mycorrhizal *Glycine max* and *Lactuca sativa* plants in relation to drought stress tolerance. Plant Molecular Biology. 2006;**60**: 389-404. DOI: 10.1007/s11103-005-4210-y

[54] Liu Z, Li Y, Ma L, Wei H, Zhang J, He X, et al. Coordinated regulation of arbuscular mycorrhizal fungi

and soybean MAPK pathway genes improved mycorrhizal soybean drought tolerance. Molecular Plant-Microbe Interactions. 2015;**28**:408-419. DOI: 10.1094/MPMI-09-14-0251-R

[55] Sharifi M, Ghorbanli M, Ebrahimzadeh H. Improved growth of salinity-stressed soybean after inoculation with salt pre-treated mycorrhizal fungi. Journal of Plant Physiology. 2007;**164**:1144-1151. DOI: 10.1016/j.jplph.2006.06.016

[56] Spagnoletti FN, Balestrasse K, Lavado RS, Giacometti R. Arbuscular mycorrhiza detoxifying response against arsenic and pathogenic fungus in soybean. Ecotoxicology and Environmental Safety. 2016;**133**:47-56. DOI: 10.1016/j.ecoenv.2016.06.012

[57] Zhang S, Zhou J, Wang G, Wang X, Liao H. The role of mycorrhizal symbiosis in aluminum and phosphorus interactions in relation to aluminum tolerance in soybean. Applied Microbiology and Biotechnology. 2015;**99**:10225-10235. DOI: 10.1007/s00253-015-6913-6

[58] Wrather A, Shannon G, Balardin R, Carregal L, Escobar R, Gupta G, et al. Effect of diseases on soybean yield in the top eight producing countries in 2006. Plant Health Progress. 2010;**10**:1094. DOI: 10.1094/PHP-2010-0125-01-RS

[59] Tylka GL, Marett CC. Distribution of the soybean cyst nematode, *Heterodera glycines*, in the United States and Canada: 1954 to 2014. Plant Health Progress. 2014;**15**(85). DOI: 10.1094/PHP-BR-14-0006

[60] Schouteden N, De Waele D, Panis B, Vos CM. Arbuscular mycorrhizal fungi for the biocontrol of plant-parasitic nematodes: A review of the mechanisms involved. Frontiers in Microbiology. 2015;**6**:1280. DOI: 10.3389/fmicb.2015.01280

[61] Tylka G, Hussey R, Roncadori R. Interactions of vesicular-arbuscular mycorrhizal fungi, phosphorus, and *Heterodera glycines* on soybean. Journal of Nematology. 1991;**23**:122

[62] Babikova Z, Gilbert L, Bruce TJA, Birkett M, Caulfield JC, Woodcock C, et al. Underground signals carried through common mycelial networks warn neighbouring plants of aphid attack. Ecology Letters. 2013;**16**:835-843. DOI: 10.1111/ele.12115

[63] Cannon SB, Ilut D, Farmer AD, Maki SL, May GD, Singer SR, et al. Polyploidy did not predate the evolution of nodulation in all legumes. PLoS One. 2010;**5**:e11630. DOI: 10.1371/journal.pone.0011630

[64] Yokota K, Hayashi M. Function and evolution of nodulation genes in legumes. Cellular and Molecular Life Sciences. 2011;**68**:1341-1351. DOI: 10.1007/s00018-011-0651-4

[65] Lindström K, Murwira M, Willems A, Altier N. The biodiversity of beneficial microbe-host mutualism: The case of rhizobia. Research in Microbiology. 2010;**161**:453-463. DOI: 10.1016/j.resmic.2010.05.005

[66] Maillet F, Poinsot V, Andre O, Puech-Pages V, Haouy A, Gueunier M, et al. Fungal lipochitooligosaccharide symbiotic signals in arbuscular mycorrhiza. Nature. 2011;**469**:58-U1501. DOI: 10.1038/nature09622

[67] Clarke VC, Loughlin PC, Day DA, Smith PMC. Transport processes of the legume symbiosome membrane. Frontiers in Plant Science. 2014;**5**. DOI: 10.3389/fpls.2014.00699

[68] Downie JA. Legume nodulation. Current Biology. 2014;**24**:R184-R190. DOI: 10.1016/j.cub.2014.01.028

[69] Oldroyd GED, Murray JD, Poole PS, Downie JA. The rules of

engagement in the legume-rhizobial symbiosis. Annual Review of Genetics. 2011;**45**:119-144. DOI: 10.1146/annurev-genet-110410-132549

[70] Fortin MG, Morrison NA, Verma DP. Nodulin-26, a peribacteroid membrane nodulin is expressed independently of the development of the peribacteroid compartment. Nucleic Acids Research. 1987;**15**:813-824. DOI: 10.1093/nar/15.2.813

[71] Oldroyd GED. Speak, friend, and enter: Signaling systems that promote beneficial symbiotic associations in plants. Nature Reviews Microbiology. 2013;**11**:252-263. DOI: 10.1038/nrmicro2990

[72] Udvardi M, Poole P. Transport and metabolism in legume-rhizobia symbioses. Annual Review of Plant Biology. 2013;**64**:781-805. DOI: 10.1146/annurev-arplant-050312-120235

[73] Collier R, Tegeder M. Soybean ureide transporters play a critical role in nodule development, function and nitrogen export. Plant Journal. 2012;**72**:355-367. DOI: 10.1111/j.1365-313X.2012.05086.x

[74] Provorov NA, Tikhonovich IA. Genetic resources for improving nitrogen fixation in legume-rhizobia symbiosis. Genetic Resources and Crop Evolution. 2003;**50**:89-99. DOI: 10.1023/a:1022957429160

[75] Gebril S, Seger M, Villanueva FM, Ortega JL, Bagga S, Sengupta-Gopalan C. Transgenic alfalfa (*Medicago sativa*) with increased sucrose phosphate synthase activity shows enhanced growth when grown under N_2-fixing conditions. Planta. 2015;**242**:1009-1024. DOI: 10.1007/s00425-015-2342-0

[76] Udvardi MK, Day DA. Metabolite transport across symbiotic membranes of legume nodules. Annual Review of Plant Physiology and Plant Molecular Biology. 1997;**48**:493-523. DOI: 10.1146/annurev.arplant.48.1.493

[77] Sugiyama A, Saida Y, Yoshimizu M, Takanashi K, Sosso D, Frommer WB, et al. Molecular characterization of LjSWEET3, a sugar transporter in nodules of *Lotus japonicus*. Plant and Cell Physiology. 2017;**58**:298-306. DOI: 10.1093/pcp/pcw190

[78] Yan J, Han XZ, Ji ZJ, Li Y, Wang ET, Xie ZH, et al. Abundance and diversity of soybean-nodulating rhizobia in black soil are impacted by land use and crop management. Applied and Environmental Microbiology. 2014;**80**:5394-5402. DOI: 10.1128/AEM.01135-14

[79] Kafle A, Garcia K, Wang X, Pfeffer PE, Strahan GD, Bucking H. Nutrient demand and fungal access to resources control the carbon allocation to the symbiotic partners in tripartite interactions of *Medicago truncatula*. Plant, Cell & Environment. 2018. DOI: 10.1111/pce.13359

[80] Harris D, Pacovsky R, Paul E. Carbon economy of soybean–rhizobium–glomus associations. New Phytologist. 1985;**101**:427-440. DOI: 10.1111/j.1469-8137.1985.tb02849.x

[81] Larimer AL, Clay K, Bever JD. Synergism and context dependency of interactions between arbuscular mycorrhizal fungi and rhizobia with a prairie legume. Ecology. 2014;**95**:1045-1054. DOI: 10.1890/13-0025.1

[82] Vadez V, Beck DP, Lasso JH, Drevon JJ. Utilization of the acetylene reduction assay to screen for tolerance of symbiotic N_2 fixation to limiting P nutrition in common bean. Physiologia Plantarum. 1997;**99**:227-232. DOI: 10.1111/j.1399-3054.1997.tb05406.x

[83] Mortimer PE, Perez-Fernandez MA, Valentine AJ. The role of arbuscular mycorrhizal colonization in the carbon

and nutrient economy of the tripartite symbiosis with nodulated *Phaseolus vulgaris*. Soil Biology & Biochemistry. 2008;**40**:1019-1027. DOI: 10.1016/j.soilbio.2007.11.014

[84] Chen BD, Li XL, Tao HQ, Christie P, Wong MH. The role of arbuscular mycorrhiza in zinc uptake by red clover growing in a calcareous soil spiked with various quantities of zinc. Chemosphere. 2003;**50**:839-846. DOI: 10.1016/s0045-6535(02)00228-x

[85] Ibiang YB, Mitsumoto H, Sakamoto K. Bradyrhizobia and arbuscular mycorrhizal fungi modulate manganese, iron, phosphorus, and polyphenols in soybean (*Glycine max* (L.) Merr.) under excess zinc. Environmental and Experimental Botany. 2017;**137**:1-13. DOI: 10.1016/j.envexpbot.2017.01.011

[86] Sakamoto K, Ogiwara N, Kaji T. Involvement of autoregulation in the interaction between rhizobial nodulation and AM fungal colonization in soybean roots. 2013;**49**:1141-1152. DOI: 10.1007/s00374-013-0804-8

[87] Kassaw T, Bridges W Jr, Frugoli J. Multiple autoregulation of nodulation (AON) signals identified through split root analysis of *Medicago truncatula sunn* and *rdn1* mutants. Plants. 2015;**4**: 209-224. DOI: 10.3390/plants4020209

[88] Meghvansi MK, Prasad K, Harwani D, Mahna SK. Response of soybean cultivars toward inoculation with three arbuscular mycorrhizal fungi and *Bradyrhizobium japonicum* in the alluvial soil. European Journal of Soil Biology. 2008;**44**:316-323. DOI: 10.1016/j.ejsobi.2008.03.003

[89] Farid M, Navabi A. N-2 fixation ability of different dry bean genotypes. Canadian Journal of Plant Science. 2015;**95**:1243-1257. DOI: 10.4141/cjps-2015-084

[90] Snellgrove RC, Splittstoesser WE, Stribley DP, Tinker PB. The distribution of carbon and the demand of the fungal symbiont in leek plants with vesicular-arbuscular mycorrhizas. New Phytologist. 1982;**92**:75-87. DOI: 10.1111/j.1469-8137.1982.tb03364.x

[91] Kleinert A, Venter M, Kossmann J, Valentine A. The reallocation of carbon in P deficient lupins affects biological nitrogen fixation. Journal of Plant Physiology. 2014;**171**:1619-1624. DOI: 10.1016/j.jplph.2014.07.017

[92] Fellbaum CR, Mensah JA, Cloos AJ, Strahan GD, Pfeffer PE, Kiers ET, et al. Fungal nutrient allocation in common mycelia networks is regulated by the carbon source strength of individual host plants. New Phytologist. 2014;**203**:645-656. DOI: 10.1111/nph.12827

[93] Fellbaum CR, Gachomo EW, Beesetty Y, Choudhari S, Strahan GD, Pfeffer PE, et al. Carbon availability triggers fungal nitrogen uptake and transport in the arbuscular mycorrhizal symbiosis. Proceedings of the National Academy of Sciences of the United States of America. 2012;**109**:2666-2671. DOI: 10.1073/pnas.1118650109

[94] Konvalinková T, Jansa J. Lights off for arbuscular mycorrhiza: On its symbiotic functioning under light deprivation. Frontiers in Plant Science. 2016;7:782. DOI: 10.3389/fpls.2016.00782

[95] Kiers ET, Duhamel M, Beesetty Y, Mensah JA, Franken O, Verbruggen E, et al. Reciprocal rewards stabilize cooperation in the mycorrhizal symbiosis. Science. 2011;**333**:880-882. DOI: 10.1126/science.1208473

[96] Afkhami ME, Stinchcombe JR. Multiple mutualist effects on genomewide expression in the tripartite association between *Medicago*

truncatula, nitrogen-fixing bacteria and mycorrhizal fungi. Molecular Ecology. 2016;**25**:4946-4962. DOI: 10.1111/mec.13809

[97] Doidy J, Van Tuinen D, Lamotte O, Corneillat M, Alcaraz G, Wipf D. The *Medicago truncatula* sucrose transporter family: Characterization and implication of key members in carbon partitioning towards arbuscular mycorrhizal fungi. Molecular Plant. 2012;**5**:1346-1358. DOI: 10.1093/mp/sss079

[98] Chen LQ. SWEET sugar transporters for phloem transport and pathogen nutrition. New Phytologist. 2014;**201**:1150-1155. DOI: 10.1111/nph.12445

[99] Kryvoruchko IS, Sinharoy S, Torres-Jerez I, Sosso D, Pislariu CI, Guan D, et al. MtSWEET11, a nodule-specific sucrose transporter of *Medicago truncatula* root nodules. Plant Physiology. 2016;**171**:554-565. DOI: 10.1104/pp.15.01910

[100] Petrini O. Fungal endophytes of tree leaves. In: Andrews JH, Hirano SS, editors. Microbial Ecology of Leaves. New York: Springer Verlag; 1991. pp. 179-197. DOI: 10.1007/978-1-4612-3168-4_9

[101] Bacon CW, White JFJ. Microbial Endophytes. New York: Marcel Dekker, Inc; 2000

[102] Impullitti AE, Malvick DK. Fungal endophyte diversity in soybean. Journal of Applied Microbiology. 2013;**114**:1500-1506. DOI: 10.1111/jam.12164

[103] Bilal S, Khan AL, Shahzad R, Asaf S, Kang S-M, Lee I-J. Endophytic *Paecilomyces formosus* LHL10 augments *Glycine max* L. adaptation to Ni-contamination through affecting endogenous phytohormones and oxidative stress. Frontiers in Plant Science. 2017;**8**:870. DOI: 10.3389/fpls.2017.00870

[104] Khan AL, Lee I-J. Endophytic *Penicillium funiculosum* LHL06 secretes gibberellin that reprograms Glycine max L. growth during copper stress. BMC Plant Biology. 2013;**13**:86. DOI: 10.1186/1471-2229-13-86

[105] Bajaj R, Hu WM, Huang YY, Chen SY, Prasad R, Varma A, et al. The beneficial root endophyte *Piriformospora indica* reduces of the soybean cyst nematode. Biological Control. 2015;**90**:193-199. DOI: 10.1016/j.biocontrol.2015.05.021

[106] Lopes KBD, Carpentieri-Pipolo V, Oro TH, Pagliosa ES, Degrassi G. Culturable endophytic bacterial communities associated with field-grown soybean. Journal of Applied Microbiology. 2016;**120**:740-755. DOI: 10.1111/jam.13046

[107] Subramanian P, Kim K, Krishnamoorthy R, Sundaram S, Sa TM. Endophytic bacteria improve nodule function and plant nitrogen in soybean on co-inoculation with *Bradyrhizobium japonicum* MN110. Plant Growth Regulation. 2015;**76**:327-332. DOI: 10.1007/s10725-014-9993-x

[108] Egamberdieva D, Jabborova D, Berg G. Synergistic interactions between *Bradyrhizobium japonicum* and the endophyte *Stenotrophomonas rhizophila* and their effects on growth, and nodulation of soybean under salt stress. Plant and Soil. 2016;**405**:35-45. DOI: 10.1007/s11104-015-2661-8

[109] De Santana AS, Tiburcio Cavalcante UM, De Sa Barreto Sampaio EV, Maia LC. Production, storage and costs of inoculum of arbuscular mycorrhizal fungi (AMF). Brazilian Journal of Botany. 2014;**37**:159-165. DOI: 10.1007/s40415-014-0056-3

[110] Douds DD Jr, Nagahashi G, Reider C, Hepperly PR. Inoculation with arbuscular mycorrhizal fungi increases the yield of potatoes in a high P soil. Biological Agriculture & Horticulture. 2007;**25**:67-78. DOI: 10.1080/01448765.2007.10823209

[111] Lojan P, Senes-Guerrero C, Pablo Suarez J, Kromann P, Schuessler A, Declerck S. Potato field-inoculation in Ecuador with *Rhizophagus irregularis*: No impact on growth performance and associated arbuscular mycorrhizal fungal communities. Symbiosis. 2017;**73**:45-56. DOI: 10.1007/s13199-016-0471-2

[112] Mensah JA, Koch AM, Antunes PM, Hart MM, Kiers ET, Bücking H. High functional diversity within arbuscular mycorrhizal fungal species is associated with differences in phosphate and nitrogen uptake and fungal phosphate metabolism. Mycorrhiza. 2015;**25**:533-546. DOI: 10.1007/s00572-015-0631-x

[113] Berruti A, Lumini E, Balestrini R, Bianciotto V. Arbuscular mycorrhizal fungi as natural biofertilizers: let's benefit from past successes. Frontiers in Microbiology. 2016;**6**:1559. DOI: 10.3389/fmicb.2015.01559

[114] Li S, Cong Y, Liu Y, Wang T, Shuai Q, Chen N, et al. Optimization of *Agrobacterium*-mediated transformation in soybean. Frontiers in Plant Science. 2017;**8**:246. DOI: 10.3389/fpls.2017.00246

[115] Sun X, Hu Z, Chen R, Jiang Q, Song G, Zhang H, et al. Targeted mutagenesis in soybean using the CRISPR-Cas9 system. Scientific Reports. 2015;**5**:10342. DOI: 10.1038/srep10342

Chapter 8

Bio-Based Epoxy Resin from Epoxidized Soybean Oil

Qiuyu Tang, Yahua Chen, Huizhi Gao, Qian Li, Zhenhao Xi, Ling Zhao, Chong Peng and Lanpeng Li

Abstract

Epoxidized soybean oil (ESO) is the oxidation product of soybean oil with hydrogen peroxide and either acetic or formic acid obtained by converting the double bonds into epoxy groups, which is non-toxic and of higher chemical reactivity. ESO is mainly used as a green plasticizer for polyvinyl chloride, while the reactive epoxy groups imply its great potential in both the monomer synthesis and the polymer preparation fields. Functional polymers are obtained by different kinds of reactions of the ESO with co-monomers and/or initiators shown in this chapter. The emphasis is on ESO based epoxy cross-linked polymers which recently gained strong interest and allowed new developments especially from both an academic point of view and an industrial point of view. It is believed that new ring-opening reagents may facilitate the synthesis of good structural ESO based materials.

Keywords: epoxidized soybean oil, cross-linking reaction, acrylated epoxidized soybean oil, polyols, polyurethanes, bio-based polymer

1. Introduction

The utilization of renewable resources in the field of polymer synthesis has gained a great deal of attention due to the growing public concerns for the environmental concerns and the sustainable development [1, 2]. Epoxidized soybean oil (ESO) is the bio-based product from the epoxidation of soybean oil with hydrogen peroxide and either acetic or formic acid obtained by converting the double bonds into epoxy groups, which is non-toxic and of higher chemical reactivity [3]. It is mainly used as a green plasticizer for many plastics currently [4]. Meanwhile it has also attracted an increasing attention as a green epoxy resin utilizing the reactive epoxy groups into both the monomer synthesis and the polymer preparation due to its low cost, little toxicity, and large production, which imply its great potential in industrial process [5].

ESO can be converted by different kinds of reactions with co-monomers and/or initiators [6]. Permanent network that comes from the directing cross-linking of ESO and hardeners endows ESO with great stability, superior mechanical properties and satisfying chemical resistance, which make the products competitive among a variety of materials. In addition, the chemical modification of ESO has gained more and more attention in recent years. Introducing hydroxyl groups to make polyols for polyurethanes synthesis is one of the most important chemical modification methods [7]. Acrylated epoxidized soybean oil (AESO) obtained by ring opening esterification between acrylic acid and ESO is of high reactivity for thermal and UV initiated polymerization [8, 9]. This chapter reviews the applications of ESO and its derivatives for the preparation of a series of bio-based polymeric materials.

2. Direct cross-linking

2.1 Amine hardeners

Functional amines are widely used as curing agents for generating epoxy resin. For ESO, a series of amines used as curing agents are listed in **Table 1** and the reaction scheme between ESO and amine is shown in **Figure 1**. Most of the researchers focused on the investigation of the cross-linking process of partially bio-based polymers because of the unsatisfying properties of fully bio-based ones. Three main methods can be applied to improve the properties of ESO-based thermosets, which are using commercial curing agents, adding commercial epoxy resins to ESO, and adding other materials to make composites [10–12].

The curing processes of ESO or the mixture of ESO and commercial epoxy resin have been investigated, and some of these systems have been made into composites through adding fibers [10–12, 14], clay [16, 18] and other reinforcement [19]. Viscoelastic properties, mechanical properties and many other analyses have been studied to evaluate their applicability to be used in industry. The partially bio-based polymers show great potential to replace fully petroleum-based polymers in many areas according to the testing results. Glass-transition (Tg) and viscoelastic properties of amine-cured ESO can be enhanced by increasing the amount of triethylenetetramine (TETA) or triethylene glycol diamine (TGD). TETA endows the polymer with similar viscoelastic properties to a commercial rubber and a higher Tg than TGD does [13]. In this respect, the biopolymers made from ESO and amines have

No.	Epoxy resin	Hardener
1	ESO	Triethylene glycol diamine (TGD) [13, 14]
2	ESO	Triethylenetetramine (TETA) [10, 13, 15–18]
3	ESO	Diethylenetriamine (DETA) [10, 15]
4	ESO	Jeffamine D-230 [10]
5	ESO	Jeffamine T-403 [10]
6	ESO	Jeffamine EDR-148 [10]
7	ESO + diglycidyl ether of bisphenol A (DGEBA)	TETA [11, 12, 19]
8	ESO + DGEBA	DETA [11, 12]
9	ESO + DGEBA	Jeffamine D-230 [11]
10	ESO + DGEBA	Jeffamine T-403 [11]
11	ESO + DGEBA	Jeffamine EDR-148 [11]
12	ESO + DGEBA	Linear polyethylenimine [12]
13	ESO, ESO + DGEBA	Dicyandiamide (DICY) [20]
14	ESO [21]	Decamethylene diamine, succinic anhydride
15	ESO + DGEBA	Isophorone diamine(IPDA) [22]

Table 1.
Amines for curing ESO and ESO composites.

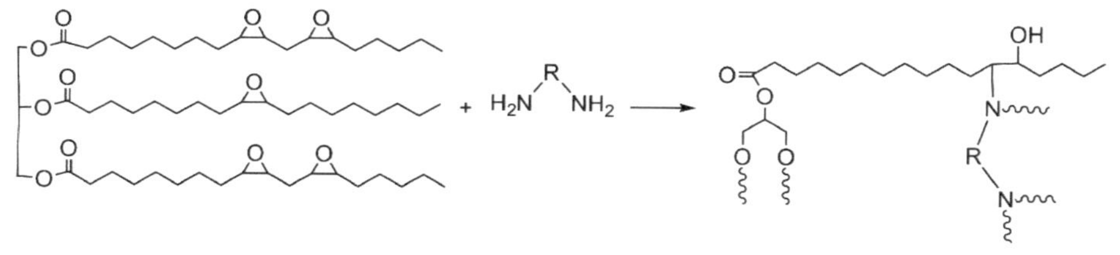

Figure 1.
The curing process between ESO and amine curing agent [17].

great potential to replace some synthetic rubbers or plastics [13, 14]. Besides, the quasi-static and dynamic compressive properties of the cured products based on ESO and amines and the corresponding composites reinforced by clay have also been investigated to develop compressive one-dimensional stress-strain material models [15, 16]. Solid freeform fabrication method has been applied to the preparation of ESO-based composites and proved to be a suitable method for this kind of curing system [10–12]. ESO/TETA/clay composites show controllable biodegradability, low cost, good thermal and mechanical properties, and these properties indicates that the composites may work as alternative to petroleum-based polymers in the field of insulation materials and coating materials [18]. For clay-reinforced composites based on commercial epoxy resin the addition of ESO can enhance the impact strengths [22]. More interestingly, the product from ESO and TETA can be made into an ion-exchange resin through hydrolysis [17]. Usually, epoxy groups in the internal of the long aliphatic chain exhibits much poorer reactivity than those terminal epoxy groups. Due to this fact, the reported curing processes of ESO usually needs higher temperature and longer time than commercial petroleum-based epoxy resin, such as bisphenol A epoxy resin. However, the combination of the hardener, dicyandiamide (DICY), and the accelerator, carbonyldiimidazole (CDI), can make the gelation of ESO occur within 13 min at 190°C [20]. Moreover, the gelation of the mixture of ESO and DGEBA is achieved with the aid of DICY and CDI within 3 min at 160°C [20].

Fully or high bio-based polymers are also attractive to researchers owing to people's strong attention to environment concerns. A series of fully bio-based elastomers have been synthesized through the ring-opening reaction between ESO and a bio-based amine hardener, decamethylene diamine, and they can be cross-linked by further reaction with another bio-based anhydride hardener, succinic anhydride [21]. These fully bio-based elastomers have great potential to replace some petroleum-based rubbers in engineering because of their good damping property, low water absorption and weak degradability in phosphate buffer solution [21].

2.2 Anhydride and acid hardeners

Anhydrides, which are less toxic than amines, are another kind of mainly-used hardeners (**Table 2**). The structure of anhydride-cured ESO is shown in **Figure 2**.

The investigation of green anhydride curing agents is one of the research priorities. Maleopimaric acid (MPA), which comes from rosin acid, has been used for ESO curing to obtain new polymeric thermosets with a high bio-based content [24, 25]. The total heat release is only 31.7 kJ/mol epoxy group. Compared with its petroleum-based analogues, MPA endows the polymer with larger breaking elongation, higher storage modulus and better thermal stability. Sebacic acid is another bio-based curing agent for ESO in lab. A fully bio-based composite with highly improved thermal and mechanical properties can be produced through interaction between sebacic-cured ESO and PLA [26, 27]. What's more, sebacic acid-cured ESO

No.	Epoxy resin	Hardener
1	ESO	Maleopimaric acid (MPA) [24, 25]
2	ESO	Methyltetrahydrophthalic anhydride (MTHPA) [34, 35]
3	ESO + DGEBA	MTHPA [22, 36]
4	ESO [37–40], ESO + DGEBA [41, 42]	Methylhexahydropthalic anhydride (MHHPA)
6	ESO	Maleic anhydride (MAL) [43, 44]
7	ESO	Phthalic anhydride [45]
8	ESO	Nadic methyl anhydride [46]
9	ESO	maleinized polybutadiene (MMPBD) [47]
10	ESO [30]	terpene-based acid anhydride (TPAn), maleinated linseed oil, hexahydrophthalic anhydride
11	ESO [44]	hexahydrophthalic anhydride (CH), MAL, succinic anhydride (SUC), dodecenylsuccinic anhydride (DDS)
12	ESO [26–28, 33], ESO + DGEBA [48]	Sebacic acid
13	ESO [33]	Adipic acid, 1,12-dodecanedicarboxylic acid, sebacic acid
14	ESO [32]	Citric acid, carboxylic acid functionalized MWCNTs
15	ESO [49], ESO + epoxidized linseed oil (ELO) [50]	Carboxyl-terminated polyester
16	ESO	Dicarboxyl terminated oligomeric poly(butylene succinate) [51]
17	ESO	Dicarboxyl-terminated polymide1010 oligomers [23]
18	ESO + ELO	Phosphorylated castor oil [31]

Table 2.
Anhydride and acid for curing ESO and ESO composites.

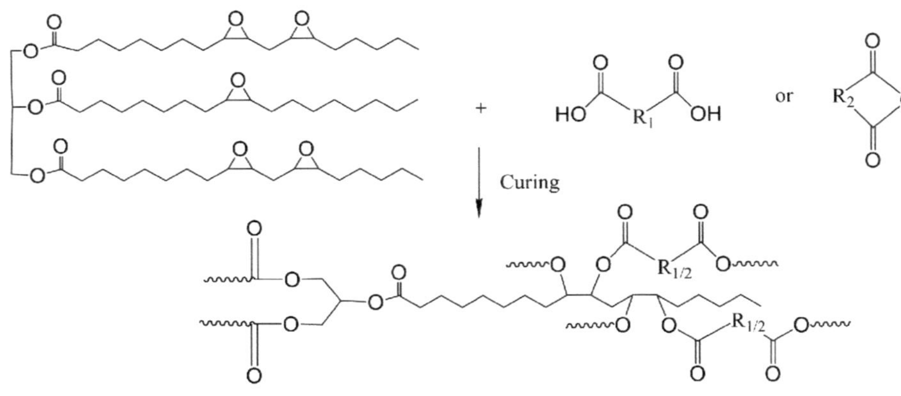

Figure 2.
The curing process between ESO and dicarboxylic acids or anhydrides [23].

can be applied in the field of superhydrophobic materials to make a sustainable and biodegradable superhydrophobic material [28, 29]. Other bio-based chemicals, such as terpene [30], vegetable oils [30, 31] and citric acid [32], are all the optional raw material for green curing agents. A terpene-based acid anhydride has been found to endow ESO with higher Tg, higher tensile strength and greater modulus than maleinated linseed oil and hexahydrophthalic anhydride do [30]. But maleinated linseed oil makes the thermoset easier to biodegrade [30]. Biodegradable and biocompatible elastomers, which may be competitive in the field of implantable materials, can be obtained by curing ESO and Epoxidized linseed oil (ELO) with phosphorylated castor oil [31]. Carboxylic acid functionalized MWCNTs are always used as the filler for fully bio-based ESO/citric acid system [32]. The produced composites with good mechanical properties and high bio-based content may be applied in the field of industry [32]. Physical tests of fully sustainable polymers obtained from curing ESO with different dicarboxylic acids show the decreases of Tg and elongation at break, and the increases of tensile strength and Young's modulus with the increasing of chain-length of the curing agents [33]. In this respect, besides bio-based micromolecular chemicals, bio-based dicarboxyl-terminated polymers are also able to work as green curing agents for ESO to make fully bio-based polymers [23]. Polymer curing agents with long chain length can avoid the short, brittle and amorphous cross-link structures which may be the reason for the poor performance of ESO-based thermosets [23].

Like the situation occurring in amine-cured systems, anhydride-cured ESO with a high bio-based content usually cannot exhibit excellent properties as petroleum-based polymers do. In order to overcome this deficiency, ESO usually works together with some petroleum-based chemicals. For this kind of complicated reaction systems, many factors are worth investigations. We are going to discuss this kind of reaction systems in terms of the properties of epoxides, the addition of commercial curing agents, the influence of the catalysts and the incorporation of fillers.

The internal epoxy rings in ESO exhibits lower reactivity than terminal ones do and the epoxy equivalent weight of ESO is usually higher than commercial epoxy resins. The addition of ESO in the mixture of DGEBA and ESO results in the increase of peak exothermic temperature, and activation energy and the decrease of enthalpy of reaction [36, 48]. Tensile strength, modulus, fracture toughness, impact strength, storage modulus (E′) in the glassy state and Tg of the cured products decrease because of the addition of ESO [36, 41]. Besides, the thermal and mechanical properties of the cured products has a positive correlation with the epoxide content of ESO [35].

Aside from the alteration of epoxides, the properties of the cured products can be enhanced with the aid of commercial curing agents. Bio-based foams based on methyl-tetrahydrophthalic anhydride (MTHPA)-cured ESO show similar mechanical properties to synthetic epoxy foams and the contents of ESO can be larger than 55 wt%, which indicates that this kind of green foams can be valuable alternative for commercial epoxy foams [34]. Polymers with anhydride groups [47] and dicarboxylic acids [49–51] are also able to work as curing agents for ESO. The carboxylic acid-terminated polyesters can work with ESO to produce green pressure-sensitive adhesives, which are environmentally friendly [50], thermal stable and with flame retardance [49]. In this kind of curing systems, the molecular weight of the polymer curing agents obviously have a great influence on the curing process and the physical properties of the cured bio-based products [51]. One of the remarkable advantages of bio-based polymers is their potential biodegradability. Lower crosslink density usually means higher biodegradability for ESO-based thermosets [40]. The cross-link density of the cured product reaches maximum at stoichiometric ratio between ESO and hardener [45].

Not only the properties of the main reactants, but the loading and type of the catalyst have a great influence on the on the curing process [38] final polymers [39]. The curing kinetics of ESO/methyl hexahydrophthalic anhydride (MHHPA) system show a significantly autocatalytic characteristic and ESO with 1.5 phr (parts per hundreds of resin) of 2-ethyl-4-methylimidazole (EMI) catalyst is a recommended composition for ESO/MHHPA system to be cured effectively at relative low temperature and short time [38].

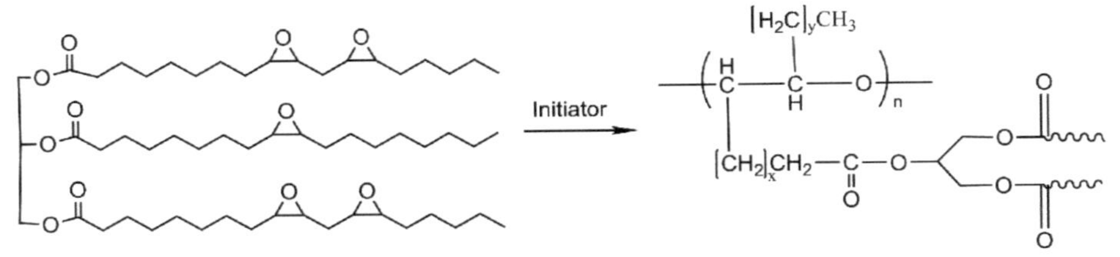

Figure 3.
Chain-growth polymerization of ESO under initiators [52].

ESO-based thermosets can also be used as good matrixes for organoclays [22, 35], organo-montmorillonite clay [37], proteins [46], regenerated cellulose [30] and other fillers. These works show that the thermal and mechanical properties of the composites can be improved significantly with the addition of different fillers.

2.3 Initiators for chain-growth polymerization

Besides adding curing agents, ESO can also be cross-linked only by initiators, as shown in **Figure 3**. Fluoroantimonic acid hexahydrate ($HSbF_6{\cdot}6H_2O$) [6] and boron trifluoride diethyl etherate ($BF_3{\cdot}OEt_2$) [52–54] are commonly employed to initiate the ring-opening polymerization of ESO. As the special macromolecular structure and mechanical properties, the products have the potential to be made into hydrogels and applied in the areas of personal and health care [6, 53]. Besides, the cross-linked ESO initiated by $BF_3{\cdot}OEt_2$ can be used to synthesize bio-based surfactants, which can help produce microbubbles effectively [54] and may take the place of petroleum-based detergents and surfactants [55].

3. Introducing hydroxyl groups

Besides the curing, introducing hydroxyl groups is one of the most important chemical modification of ESO. Hydroxyl groups are functional groups that can be compatible with matrixes through hydrogen bonding or can be able to covalently bond with matrixes using some active chemicals [56].

3.1 ESO-based polyols

Bio-based polyols with two or more hydroxyl groups can be synthesized from ESO by epoxy ring opening applying different approaches (see **Figure 4**). Ring opening reagents mainly include in mono-functional amines, alcohols (such as methanol, ethylene glycol, propylene glycol or butanol), acids (such as acrylic acid, acetic acid, phosphoric acid, fatty acids, carboxylic acid, hexanoic acids, or octanoic acids), thioethers or ketones [57–64]. Lewis acid is known as a kind of useful initiator for the hydroxyl reaction with epoxides. ESO-based polyether polyols are capable to be prepared by Lewis acids catalyzing ring opening with propylene glycol [60]. After that, the ESO-based polyether polyols with higher molecular weight can be cured with phenolic, melamine and other conventional crosslinkers to give reasonable film properties [65]. Besides, ESO phosphate ester polyols have been synthesized by using super phosphoric acid phosphorylated ESO, which is able to be incorporated in bake coatings with excellent performance [62]. A series of methoxylated soybean oil polyols (MSOLs) have been prepared with different hydroxyl functionalities by the ring opening of ESO with methanol [66]. These polyols have been applied to synthesize the environmentally friendly vegetable-oil-based

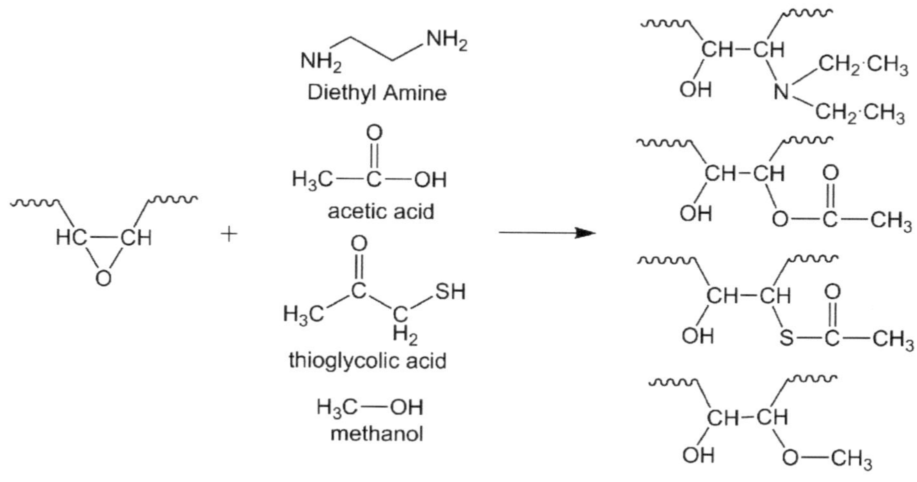

Figure 4.
Epoxy ring opening reactions with various ring opening reagents [57].

polyurethane dispersions (PUDs) with very promising properties. Thioglycolic acid (TGA) bearing thiol and carboxylic acid as two different functional groups, glycolic acid (GA) containing hydroxyl and carboxyl functionality and methyl ester of thioglycolic acid (TGAME) have been also used as ring opening agents of ESO to synthesize novel bio-based polyols [57, 67]. Using TGA and GA, the epoxy rings are opened by the carboxylic acid group, while the epoxy rings are opened by the thiol group primarily when using TGAME. In addition, polyols obtained by ring opening with TGA have higher molecular weight comparing to GA and TGAME. That is because some of the thiol groups of TGA initially remain intact and then are involved in ring opening of other epoxy groups resulting in chain coupling [57, 67].

There are some side reactions occurring during the ring-opening of ESO epoxide groups, and these side reactions often depend on reaction parameters [68, 69]. A substantial degree of oligomerization due to oxirane-oxirane, and oxirane-hydroxyl reaction will take place in the presence of phosphoric acid. It is possible to synthesize ESO-based polyols having varying hydroxyl content and phosphate-ester functionality by controlling the type and amount of polar solvent and phosphoric acid content [70]. Inter-esterification or intermolecular ether formation are also observed as side reactions, depending on the molar proportion of the hydrogen donor [68]. Different catalysts for the ring opening of the epoxide groups in ESO have been evaluated in many works. The most common catalysts are sulfuric acid, p-toluenesulfonic acid, perchloric acid, tetrafluoroboric acid (HBF_4) and activated clays. HBF_4 have been found to produce polyols with a higher OH content, and lower viscosity than other catalysts in the ring opening reaction of ESO with methanol [69]. And, triflic acid is a very effective catalyst for preparing ESO polyether polyols [60]. As alcohol concentration relative to ESO is reduced, higher molecular weight polyether polyols can be produced in a controlled way [60].

3.2 ESO-based polyurethanes

Currently, vegetable oils-based polyols are gradually replacing petroleum-based hydroxyl for preparing PUs, which are considered as sustainable and environmentally friendly polymers from biomass industry [5]. ESO based polyols can be co-polymerized with some commercial isocyanates, such as toluene di-isocyanate (TDI), methylene-4,49-diphenyldiisocyanate (MDI) or others, to obtain

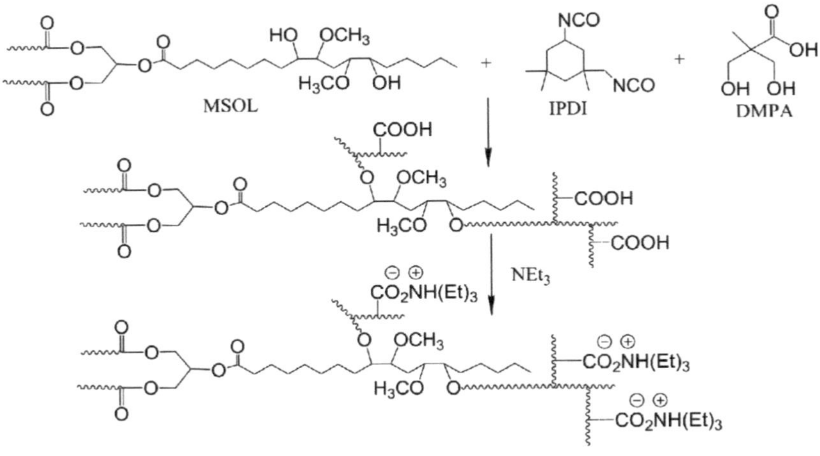

Figure 5.
Synthesis of soybean-oil-based PUs [66].

bio-based PUs with useful properties, including enhanced hydrolytic and thermal stability, as shown in **Figure 5**.

The structure-property relationships between ESO based polyols and PUs have been extensively investigated. Several factors have important influences on the properties of the PUs, such as chemical structure of the segment, chemical composition, hydroxyl group position, hydroxyl values of polyols and cross-linking densities of the PUs networks [71]. The structure and properties of PUs prepared from halogenated as well as non-halogenated soybean polyols with commercial isocyanates have been studied which shows that brominated polyols and their corresponding PUs have the highest densities and Tg while their thermal stabilities are lowest. Chlorinated polyols have comparable glass transition and strength to brominated polyols, somewhat higher than the methoxy-containing and hydrogenated polyols [69]. Besides, the NCO/OH mole ratios also show effects on the properties of the PUs networks that the cross-linking densities, Tg, and tensile strengths deteriorate as the NCO/OH ratios decrease and glassy polymers can be produced when the NCO/OH ratio is between 0.8 and 1.05 [72]. The studies on polyurethane resins from a blend of glycerol and polyol show that the increasing of Tg caused by the incorporation of glycerol into soy polyols obviously enhances the rigidity of PUs [73]. The polyurethanes elastomers synthesized from ESO based polyols obtained by ring opening with Ricinoleic acid (RA) and sebacic acid with citric acid as the cross-linker display biocompatibility and biodegradability and are very suitable for bone tissue engineering [74].

Furthermore, ESO is able to be effectively converted to carbonated soybean oil (CSBO) containing five-membered cyclic carbonates by reacting with carbon dioxide in the presence of tetra-butylammonium bromide at 110°C in high yield [75]. Then, CSBO can easily react with diamines to give the corresponding non-isocyanate polyurethane networks(NIPUs), and the thermal and mechanical properties of NIPUs can be well adjusted and controlled by changing the CSBO/amine ratio [76].

4. Acrylated epoxidized soybean oil (AESO)

4.1 Synthesis of AESO

AESO is commercially-manufactured derivative of ESO and has been extensively used in coatings, resins and composites. The acid-catalyzed synthesis process of AESO is shown in **Figure 6**. The acid catalyst promotes the formation of an oxonium ion, which can be stabilized by local epoxide group. And the ring-opening reaction happened between acrylic acid and the oxonium ion. Inhibitor is needed in this reaction to prevent polymerization of vinyl groups. The acrylation reaction has a first-order dependence on the concentration of epoxy groups, but the rate constant increases with the decreasing of epoxides per fatty acid due to steric hindrance and the stabilization effect of local epoxide group on oxonium groups [77].

4.2 Thermal initiation of AESO

Through reversible addition-fragmentation chain transfer (RAFT) polymerization, AESO can be made into a hyper-branched bio-based polymer without macrogelation [8, 79]. The conversion of vinyl is usually over 50%, which indicates that it is possible for multifunctional renewable feed stocks to be made into bio-based thermoplastics polymers at a high conversion without gelation [8].

Most of the researches focused on the cross-linking reaction of AESO through free radical polymerization. Like the ESO, the cross-linked homopolymers from AESO also have the shortage that the polymers exhibit poor mechanical properties [80]. One of the common methods used to enhance its mechanical properties is adding reinforcements to make polymer composites. There are many polar groups in the structure of AESO, including C=O, —OH and epoxy groups. These polar groups provide the possibility for the formation of hydrogen bonds between AESO and fillers [80]. Thermoplastic polyurethane [81], microcrystalline cellulose (MCC) [80] and cellulose fiber [82] are the common reinforcements worth investigation for poly(acrylated epoxidized soybean oil)(PAESO). The interaction between PAESO and polyurethane can be enhanced by the formation of hydrogen bonds between hydrophilic functional groups from both of the two components which give rise to the result of improving the toughness and increasing the elongation of PAESO [81]. As a green filler, microcrystalline cellulose will increase the density, hardness, flexural strength and modulus of the material without decreasing the bio-based content [80]. Cellulose-reinforced PAESO can also be successfully made into bio-based foams with enhanced mechanical properties, which shows the great potential to replace petroleum-based foams [82].

Another common way to adjust the properties of AESO-based materials is the incorporation of co-monomers. Styrene [83–86], N-vinyl-2-pyrrolidone (NVP) [64, 87], 3-isopropenyldimethylbenzyl isocyanate (TMI) [88], isocyanatoethyl methacrylate (IEM) [88], 1,6-hexanediol diacrylate [89], divinylbenzene [86, 89] and unsaturated polyester [90–94] are widely used as co-monomers for AESO. The diblock copolymers based on AESO and styrene are able to work as an additive for asphalt to modify the rheological performance so that the corresponding stiffness, elasticity and rutting resistance of the asphalt can be substantially improved [83]. The copolymer based on AESO and styrene can also be reinforced by natural fibers

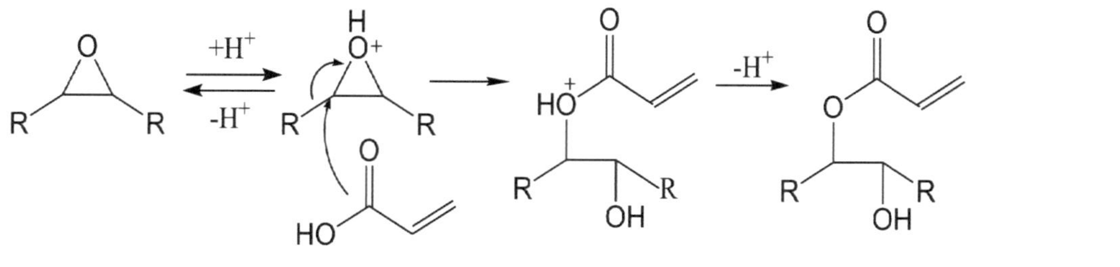

Figure 6.
Mechanism of AESO synthesis [78].

[84] and denim [85] to obtain bio-based composites for structural applications, such as roof structure and safety helmets. Due to the toxicity of styrene, styrene-free polymers become more attractive recently. NVP is an alternative to styrene in the synthesis of copolymer based on AESO, and the corresponding hemp fibers (HFs) composites exhibit superior static and dynamic mechanical properties [64]. As both AESO and HFs contains —OH groups in their structures, the addition of isophorone diisocyanate, whose isocyanate groups can react with —OH groups, to the AESO/HFs/NVP system can improve the properties by working as both a cross-linker and a coupling agent [87]. Accordingly, TMI and IEM bringing both C=C double bonds and isocyanate groups into the reaction systems may also be good co-monomers for AESO/HFs system. Besides the free radical polymerization of vinyl groups, the reactions between isocyanate groups and the —OH groups of AESO and HFs also occurred at the same time in this bio-based polymer composite systems [88]. Consequently, the crosslinking density and interfacial reaction between reinforcement and the matrix can be improved significantly, leading to the enhancement of storage modulus, Tg and water resistance. As a nonvolatile and nonhazardous chemical, AESO is a suitable replacement for styrene in unsaturated polyester (UPE) resin to obtain hybrid polymer networks [90–94]. The UPE with unsaturated sites works as the co-monomer for AESO, and the final products usually exhibit comparable properties to correspondingly styrene-based products. The combination of a variety of co-monomers may provide AESO based copolymers with more possibilities. The thermosets based on the combination of AESO, styrene and divinylbenzene can be the potential replacements for commercial electronic materials [86]. The combination of AESO, 1,6-hexanediol diacrylate and divinylbenzene is able to make into the matrix for bacterial cellulose nanocomposite foams and the properties of the composites can be tailored by adjusting the compositions [89].

Although petroleum-based co-monomers can bring excellent properties, the decrease of the bio-based content is still not expected. Functional bio-based co-monomers are desired in consequence. Isosorbide can be used to synthesize a bio-based co-monomer for AESO through the reaction with methacrylate anhydride [95]. The product, isosorbide-methacrylate (IM), which has stiff structure, endows the bio-based networks with ideal thermal and mechanical properties. Similarly, rosin is also a bio-based raw material with a rigid molecular structure. Its derivative, N-dehydroabietic acrylamide (DHA-AM), can enhance the storage modulus, Tg, thermal stability, tensile strength and hydrophobicity of AESO/DHA-AM thermosets [96]. Methacrylated lauric acid (MLAU) is another bio-based reactive diluent for AESO. The mixture exhibits a suitable viscosity for liquid molding techniques to get AESO based thermoset specimens with low densities and Tg around room temperature [97].

4.3 UV curing of AESO

AESO has been widely applied in the UV curing systems for their lower volatility and relatively higher reactivity of C=C bonds which are able to conduct free-radical polymerization in the presence of functional initiator. In general, residual internal stress in the UV-curing coating film often leads to poor adhesion with substrate. AESO can be used to synthesize cured films with reduced internal stress and its flexible triglyceride structure can improve adhesion [9]. UV-curable materials based on AESO have been found many applications like coatings, adhesives and composite materials [98]. As petroleum-based fiber composites often swell after water absorption resulting in deterioration of mechanical properties, the dried distillers grains (DDGS)-flax mat coated with AESO polymerized by UV light with the initiation of irgacure 819 shows improved water resistance property [99].

Besides, AESO-based UV-cured PUDs with higher functionality can be used in textiles printing. Different content of AESO based UV-curable PUDs pigment prints adhesive have been successfully synthesized with isophorone diisocyanate (IPDI), poly(caprolactone glycol) and 2-hydroxyethyl methacrylate, and all UV-curing films have excellent thermal stability [98]. With the increasing of AESO content, the color strength of printed fabrics can be enhanced correspondingly. Conversely, the increasing of UV radiation time shows positive impact on the color fastness [100]. UV-curable, AESO-based organic shape-stabilized phase change materials also can be obtained by UV technique with enhanced thermal performance, decreased melting and freezing temperature, which verify the promising application of UV-curable material for thermal energy storage [100].

However, the existing of soft long aliphatic chains usually results in low mechanical or thermal properties and some rigid compounds are often added as the co-monomers to improve the performances of AESO-based UV-curable materials. Acrylate acid is one of the most common-used petroleum-based rigid compounds. The performances of AESO-based UV curable coating materials by using petroleum-based hyper-branched acrylates (HBAs) as co-photo-polymerization monomer, using acrylated sucrose (AS) as tougheners and using tetra-hydrofurfural acrylate (THFA) as reactive diluents show the increased coating hardness, adhesion, modulus, solvent resistance and glass transition temperature [101]. Nowadays, many researchers are devoted to exploit bio-based co-monomers to develop high bio-based content UV-curable coatings. Monomer acrylated betulin (AB) synthesized from botulin [102], unsaturated monomer (named IG) synthesized from itaconic acid and glycidyl methacrylate [103], monomers (named EM2G and EM3G) synthesized from eugenol via a thiol-ene reaction and epoxide ring-opening reaction [104] have been all evaluated to be successfully used with AESO matrix polymer and have great potential to improve the properties of UV curable coating. Coating films containing AB from 5 to 10 wt% contents have better modulus of elasticity, tensile strength, abrasion resistance and hardness, higher Tg and lower strain at break value, while the transmittance of the cured films is reduced with increasing AB loading, especially for wavelengths below 650 nm. In comparison, the polycyclic structure of betulin imposes a more rigid structure on AESO matrix polymer to enhance the applied performance [102]. In the presence of irgacure 184 as initiator, a series of UV-cured coatings without any solvent can be successfully prepared with IG (EM2G or EM3G) and AESO, and EM2G and EM3G show higher reactivity when copolymerized with AESO. The introduction of IG, EM2G and EM3G in the UV-curing system results in significantly improved mechanical and thermal properties as well as coating performances such as hardness, flexibility, adhesion, solvent resistance [103, 104].

5. ESO-based polymer composites

ESO is initially used as a plasticizer in industry for poly(vinyl chloride) chlorinated (PVC) rubber, and poly(vinyl alcohol) (PVA) emulsions to improve stability and flexibility [105, 106], and ESO is also considered to be potential nontoxic biocompatible plasticizers for poly(3-hydroxybutyrate) (PHB) and polylactic acid (PLA) when combined with other plasticizers [107–109]. Moreover, it is an interesting trend to prepare composites of ESO or its homo-polymers with other materials because of their special properties. A double network composites with ESO and a di-hydrocoumarin derived network can been synthesized with toughening effect, which make the ESO-based polymer possible to be applied in the fields of coatings and films [110]. The composites of cross-linked ESO and acrylic monolith [111] or poly(lactic acid) [112] apparently exhibit much larger Young's modulus and tensile strength than ESO homo-polymer and can work as shape memory materials, which makes ESO a potential component for manufacture of intelligent polymer materials.

Interestingly, the long chain alkane fatty acid residues in ESO can give the composites hydrophobicity, so cross-linked ESO can also work as a water-resistant film for paper that the obtained composites may be competitive in the field of packaging considering their good properties [113]. An efficient method has been reported for the formation of cellulose-based materials grafting with poly epoxidized soybean oil (PESO) with controllable hydrophobic properties [114] 1–2. A kind of PESO coated paper composites with good water-resistant property have been obtained via in situ polymerization of ESO on the surface of the paper cellulose fibers [113].

6. Conclusions

This chapter summarizes the most recent advances in the application of ESO and its derivatives for preparation of bio-based polymeric materials. The multiple reactive epoxy groups from triglycerides of unsaturated fatty acids imply its great potential in the bio-based polymer preparation fields with controllable biodegradability, thermal and mechanical properties. ESO can crosslink directly with variety curing agents to form permanent network, or to introducing reactive function groups by chemical modifications. Two most important modifications are introducing hydroxyl groups and esterification to produce acrylates. Based on these, varieties of new polymeric materials have been prepared recently from ESO and derivatives that exhibit industrially viable thermos-physical and mechanical properties and thus may find many possible applications. It is believed that ESO based compounds will gain continuously strong interest and allow new developments both in academic and industrial points of view.

Acknowledgements

The authors are grateful to the National Key Research and Development Program of China (Grant no. 2016YFB0302701), the National Natural Science Foundation of China (Grant no. 21676083), the Shanghai Rising-Star Program (Grant no. 16QB140130), the Fundamental Research Funds for the Central Universities, and the 111 Project (B08021).

Conflict of interest

The authors have declared that no conflict of interest exists.

Author details

Qiuyu Tang[1†], Yahua Chen[1†], Huizhi Gao[1], Qian Li[1], Zhenhao Xi[1*], Ling Zhao[1], Chong Peng[2] and Lanpeng Li[2]

1 Shanghai Key Laboratory of Multiphase Materials Chemical Engineering, East China University of Science and Technology, Shanghai, China

2 Dalian Research Institute of Petroleum and Petrochemicals, SINOPEC, Dalian, China

† Q. Tang and Y. Chen contributed equally.

*Address all correspondence to: zhhxi@ecust.edu.cn

References

[1] Miao S, Wang P, Su Z, Zhang S. Vegetable-oil-based polymers as future polymeric biomaterials. Acta Biomaterialia. 2014;**10**:1692-1704. DOI: 10.1016/j.actbio.2013.08.040

[2] Auvergne R, Caillol S, David G, Boutevin B, Pascault JP. Biobased thermosetting epoxy: Present and future. Chemical Reviews. 2014;**114**:1082-1115. DOI: 10.1021/cr3001274

[3] Gerbase AE, Gregório JR, Martinelli M, Brasil MC, Mendes ANF. Epoxidation of soybean oil by the methyltrioxorhenium-CH_2Cl_2/H_2O_2 catalytic biphasic system. Journal of the American Oil Chemists Society. 2002;**79**:179-181. DOI: 10.1007/s11746-002-0455-0

[4] Bueno-Ferrer C, Garrigós MC, Jiménez A. Characterization and thermal stability of poly(vinyl chloride) plasticized with epoxidized soybean oil for food packaging. Polymer Degradation and Stability. 2010;**95**:2207-2212. DOI: 10.1016/j.polymdegradstab. 2010.01.027

[5] Meier MR, Metzger JO, Schubert US. Plant oil renewable resources as green alternatives in polymer science. Chemical Society Reviews. 2007;**36**:1788. DOI: 10.1039/b703294c

[6] Liu Z, Biswas A. Fluoroantimonic acid hexahydrate ($HSbF_6{\cdot}6H_2O$) catalysis: The ring-opening polymerization of epoxidized soybean oil. Applied Catalysis A: General. 2013;**453**:370-375. DOI: 10.1016/j.apcata.2012.12.028

[7] Liu W, Xie T, Qiu R. Bamboo fibers grafted with a soybean-oil-based monomer for its unsaturated polyester composites. Cellulose. 2016;**23**:2501-2513. DOI: 10.1007/s10570-016-0969-z

[8] Yan M, Huang Y, Lu M, Lin F-Y, Hernandez NB, Cochran EW. Gel point suppression in RAFT polymerization of pure acrylic cross-linker derived from soybean oil. Biomacromolecules. 2016;**17**:2701-2709. DOI: 10.1021/acs.biomac.6b00745

[9] Rengasamy S, Mannari V. Development of soy-based UV-curable acrylate oligomers and study of their film properties. Progress in Organic Coatings. 2013;**76**:78-85. DOI: 10.1016/j.porgcoat.2012.08.012

[10] Liu ZS, Erhan SZ, Xu J, Calvert PD. Development of soybean oil-based composites by solid freeform fabrication method: Epoxidized soybean oil with bis or polyalkyleneamine curing agents system. Journal of Applied Polymer Science. 2002;**85**:2100-2107. DOI: 10.1002/app.10763

[11] Liu ZS, Erhan SZ, Calvert PD. Solid freeform fabrication of epoxidized soybean oil/epoxy composite with bis or polyalkyleneamine curing agents. Composites Part A: Applied Science and Manufacturing. 2007;**38**:87-93. DOI: 10.1016/j.compositesa.2006.01.009

[12] Liu ZS, Erhan SZ, Calvert PD. Solid freeform fabrication of epoxidized soybean oil/epoxy composites with di-, tri-, and polyethylene amine curing agents. Journal of Applied Polymer Science. 2004;**93**:356-363. DOI: 10.1002/app.20412

[13] Xu J, Liu Z, Erhan SZ, Carriere CJ. Cross-linkers control the viscoelastic properties of soybean oil-based biomaterials. Journal of the American Oil Chemists' Society. 2004;**81**:813-816. DOI: 10.1007/s11746-004-0984-6

[14] Xu J, Liu Z, Erhan SZ, Carriere CJ. A potential biodegradable rubber—Viscoelastic properties of a soybean oil-based composite. Journal

of the American Oil Chemists' Society. 2002;**79**:593-596. DOI: 10.1007/s11746-002-0528-0

[15] Song B, Chen W, Liu Z, Erhan SZ. Compressive properties of soybean oil-based polymers at quasi-static and dynamic strain rates. Journal of Applied Polymer Science. 2006;**99**:2759-2770. DOI: 10.1002/app.22627

[16] Song B, Chen W, Liu Z, Erhan S. Compressive properties of epoxidized soybean oil/clay nanocomposites. International Journal of Plasticity. 2006;**22**:1549-1568. DOI: 10.1016/j.ijplas.2005.10.004

[17] Liu ZS, Erhan SZ. Conversion of soybean oil into ion exchange resins: Removal of copper (II), nickel (II), and cobalt (II) ions from dilute aqueous solution using carboxylate-containing resin. Journal of Applied Polymer Science. 2002;**84**:2386-2396. DOI: 10.1002/app.10504

[18] Liu Z, Erhan SZ, Xu J. Preparation, characterization and mechanical properties of epoxidized soybean oil/clay nanocomposites. Polymer. 2005;**46**:10119-10127. DOI: 10.1016/j.polymer.2005.08.065

[19] Bhosale SH, Singh VV, Rangasai MC, Bandyopadhyay-Ghosh S, Ghosh SB. Integration of biobased functionalized feedstock and plastisol in epoxy resin matrix toward developing structural jute biocomposites with enhanced impact strength and moisture resistance properties. Polymer Composites. 2016;**37**:391-397. DOI: 10.1002/pc.23192

[20] Zhao CH, Wan SJ, Wang L, Liu XD, Endo T. Carbonyldiimidazole-accelerated efficient cure of epoxidized soybean oil with dicyandiamide. Journal of Polymer Science Part A: Polymer Chemistry. 2014;**52**:375-382. DOI: 10.1002/pola.27010

[21] Wang Z, Zhang X, Wang R, Kang H, Qiao B, Ma J, et al. Synthesis and characterization of novel soybean-oil-based elastomers with favorable Processability and tunable properties. Macromolecules. 2012;**45**:9010-9019. DOI: 10.1021/ma301938a

[22] Salam H, Dong Y, Davies IJ, Pramanik A. Optimization of material formulation and processing parameters in relation to mechanical properties of bioepoxy/clay nanocomposites using Taguchi design of experiments. Journal of Applied Polymer Science. 2018;**135**:45769. DOI: 10.1002/app.45769

[23] Jian X-Y, An X-P, Li Y-D, Chen J-H, Wang M, Zeng J-B. All plant oil derived epoxy thermosets with excellent comprehensive properties. Macromolecules. 2017;**50**:5729-5738. DOI: 10.1021/acs.macromol.7b01068

[24] Chen Y, Xi Z, Zhao L. New bio-based polymeric thermosets synthesized by ring-opening polymerization of epoxidized soybean oil with a green curing agent. European Polymer Journal. 2016;**84**:435-447. DOI: 10.1016/j.eurpolymj.2016.08.038

[25] Chen Y, Xi Z, Zhao L. Curing kinetics of bio-based epoxy resin based on epoxidized soybean oil and green curing agent. AICHE Journal. 2017;**63**:147-153. DOI: 10.1002/aic.15486

[26] Zhao T-H, Wu Y, Li Y-D, Wang M, Zeng J-B. High performance and thermal processable dicarboxylic acid cured epoxidized plant oil resins through dynamic vulcanization with poly(lactic acid). ACS Sustainable Chemistry and Engineering. 2017;**5**:1938-1947. DOI: 10.1021/acssuschemeng.6b02684

[27] Zhao T-H, Yuan W-Q, Li Y-D, Weng Y-X, Zeng J-B. Relating chemical structure to toughness via morphology control in fully sustainable sebacic acid cured epoxidized soybean oil toughened

polylactide blends. Macromolecules. 2018;**51**:2027-2037. DOI: 10.1021/acs.macromol.8b00103

[28] Cheng Q-Y, An X-P, Li Y-D, Huang C-L, Zeng J-B. Sustainable and biodegradable superhydrophobic coating from epoxidized soybean oil and ZnO nanoparticles on cellulosic substrates for efficient oil/water separation. ACS Sustainable Chemistry and Engineering. 2017;**5**:11440-11450. DOI: 10.1021/acssuschemeng.7b02549

[29] Cheng Q-Y, Liu M-C, Li Y-D, Zhu J, Du A-K, Zeng J-B. Biobased super-hydrophobic coating on cotton fabric fabricated by spray-coating for efficient oil/water separation. Polymer Testing. 2018;**66**:41-47. DOI: 10.1016/j.polymertesting.2018.01.005

[30] Takahashi T, Hirayama K-I, Teramoto N, Shibata M. Biocomposites composed of epoxidized soybean oil cured with terpene-based acid anhydride and cellulose fibers. Journal of Applied Polymer Science. 2008;**108**:1596-1602. DOI: 10.1002/app.27866

[31] Liu Z, Xu Y, Cao L, Bao C, Sun H, Wang L, et al. Phosphoester cross-linked vegetable oil to construct a biodegradable and biocompatible elastomer. Soft Matter. 2012;**8**:5888. DOI: 10.1039/c2sm25115a

[32] Gogoi P, Horo H, Khannam M, Dolui SK. In situ synthesis of green bionanocomposites based on aqueous citric acid cured epoxidized soybean oil-carboxylic acid functionalized multiwalled carbon nanotubes. Industrial Crops and Products. 2015;**76**:346-354. DOI: 10.1016/j.indcrop.2015.06.057

[33] Zeng R-T, Wu Y, Li Y-D, Wang M, Zeng J-B. Curing behavior of epoxidized soybean oil with biobased dicarboxylic acids. Polymer Testing. 2017;**57**:281-287. DOI: 10.1016/j.polymertesting.2016.12.007

[34] Altuna FI, Ruseckaite RA, Stefani PM. Biobased thermosetting epoxy foams: Mechanical and thermal characterization. ACS Sustainable Chemistry and Engineering. 2015;**3**:1406-1411. DOI: 10.1021/acssuschemeng.5b00114

[35] Tanrattanakul V, Saithai P. Mechanical properties of bioplastics and bioplastic-organoclay nanocomposites prepared from epoxidized soybean oil with different epoxide contents. Journal of Applied Polymer Science. 2009;**114**:3057-3067. DOI: 10.1002/app.30842

[36] Altuna FI, Espósito LH, Ruseckaite RA, Stefani PM. Thermal and mechanical properties of anhydride-cured epoxy resins with different contents of biobased epoxidized soybean oil. Journal of Applied Polymer Science. 2011;**120**:789-798. DOI: 10.1002/app.33097

[37] Tan SG, Ahmad Z, Chow WS. Reinforcing ability and co-catalytic effects of organo-montmorillonite clay on the epoxidized soybean oil bio-thermoset. Applied Clay Science. 2014;**90**:11-17. DOI: 10.1016/j.clay.2014.01.005

[38] Tan SG, Ahmad Z, Chow WS. Relationships of cure kinetics and processing for epoxidized soybean oil bio-thermoset. Industrial Crops and Products. 2013;**43**:378-385. DOI: 10.1016/j.indcrop.2012.07.049

[39] Tan SG, Chow WS. Effects of 2-ethyl-4-methylimidazole and tetraethylammonium bromide on the fracture properties of epoxidized soybean oil based thermoset. Iranian Polymer Journal. 2012;**21**:353-363. DOI: 10.1007/s13726-012-0036-z

[40] Chow WS, Tan SG, Ahmad Z, Chia KH, Lau NS, Sudesh K. Biodegradability of epoxidized soybean oil based thermosets in compost soil

environment. Journal of Polymers and the Environment. 2013;**22**:140-147. DOI: 10.1007/s10924-013-0615-x

[41] Kumar S, Samal SK, Mohanty S, Nayak SK. Epoxidized soybean oil-based epoxy blend cured with anhydride-based cross-linker: Thermal and mechanical characterization. Industrial and Engineering Chemistry Research. 2017;**56**:687-698. DOI: 10.1021/acs.iecr.6b03879

[42] Kumar S, Samal SK, Mohanty S, Nayak SK. Study of curing kinetics of anhydride cured petroleum-based (DGEBA) epoxy resin and renewable resource based epoxidized soybean oil (ESO) systems catalyzed by 2-methylimidazole. Thermochimica Acta. 2017;**654**:112-120. DOI: 10.1016/j.tca.2017.05.016

[43] España JM, Sánchez-Nacher L, Boronat T, Fombuena V, Balart R. Properties of biobased epoxy resins from epoxidized soybean oil (ESBO) cured with maleic anhydride (MA). Journal of the American Oil Chemists' Society. 2012;**89**:2067-2075. DOI: 10.1007/s11746-012-2102-2

[44] Gerbase AE, Petzhold CL, Costa APO. Dynamic mechanical and thermal behavior of epoxy resins based on soybean oil. Journal of the American Oil Chemists' Society. 2002;**79**:797-802. DOI: 10.1007/s11746-002-0561-z

[45] Gupta AP, Ahmad S, Dev A. Development of novel bio-based soybean oil epoxy resins as a function of hardener stoichiometry. Polymer-Plastics Technology and Engineering. 2010;**49**:657-661. DOI: 10.1080/03602551003681796

[46] Fombuena V, Sánchez-Nácher L, Samper MD, Juárez D, Balart R. Study of the properties of thermoset materials derived from epoxidized soybean oil and protein fillers. Journal of the American Oil Chemists' Society. 2012;**90**:449-457. DOI: 10.1007/s11746-012-2171-2

[47] ztürk C, Küsefoğlu SH. Polymerization of epoxidized soybean oil with maleinized polybutadiene. Journal of Applied Polymer Science. 2011;**120**:116-123. DOI: 10.1002/app.33094

[48] Mashouf Roudsari G, Mohanty AK, Misra M. Study of the curing kinetics of epoxy resins with biobased hardener and epoxidized soybean oil. ACS Sustainable Chemistry and Engineering. 2014;**2**:2111-2116. DOI: 10.1021/sc500176z

[49] Wang X-L, Chen L, Wu J-N, Fu T, Wang Y-Z. Flame-retardant pressure-sensitive adhesives derived from epoxidized soybean oil and phosphorus-containing dicarboxylic acids. ACS Sustainable Chemistry and Engineering. 2017;**5**:3353-3361. DOI: 10.1021/acssuschemeng.6b03201

[50] Vendamme R, Eevers W. Sweet solution for sticky problems: Chemoreological design of self-adhesive gel materials derived from lipid biofeedstocks and adhesion tailoring via incorporation of isosorbide. Macromolecules. 2013;**46**:3395-3405. DOI: 10.1021/ma400203v

[51] Jian X-Y, He Y, Li Y-D, Wang M, Zeng J-B. Curing of epoxidized soybean oil with crystalline oligomeric poly(butylene succinate) towards high performance and sustainable epoxy resins. Chemical Engineering Journal. 2017;**326**:875-885. DOI: 10.1016/j.cej.2017.06.039

[52] Liu Z, Doll KM, Holser RA. Boron trifluoride catalyzed ring-opening polymerization of epoxidized soybean oil in liquid carbon dioxide. Green Chemistry. 2009;**11**:1774. DOI: 10.1039/b915742p

[53] Liu Z, Erhan SZ. Ring-opening polymerization of epoxidized soybean

oil. JAOCS, Journal of the American Oil Chemists' Society. 2010;**87**:437-444. DOI: 10.1007/s11746-009-1514-0

[54] Xu Q, Liu Z, Nakajima M, Ichikawa S, Nakamura N, Roy P, et al. Characterization of a soybean oil-based biosurfactant and evaluation of its ability to form microbubbles. Bioresource Technology. 2010;**101**:3711-3717. DOI: 10.1016/j.biortech.2009.12.093

[55] Biresaw G, Liu ZS, Erhan SZ. Investigation of the surface properties of polymeric soaps obtained by ring-opening polymerization of epoxidized soybean oil. Journal of Applied Polymer Science. 2008;**108**:1976-1985. DOI: 10.1002/app.27784

[56] Lu Y, Larock RC. Novel polymeric materials from vegetable oils and vinyl monomers: Preparation, properties, and applications. ChemSusChem. 2009;**2**:136-147. DOI: 10.1002/cssc.200800241

[57] Karadeniz K, Aki H, Sen MY, Calikolu Y. Ring opening of epoxidized soybean oil with compounds containing two different functional groups. JAOCS, Journal of the American Oil Chemists' Society. 2015;**92**:725-731. DOI: 10.1007/s11746-015-2638-z

[58] Caillol S, Desroches M, Boutevin G, Loubat C, Auvergne R, Boutevin B. Synthesis of new polyester polyols from epoxidized vegetable oils and biobased acids. European Journal of Lipid Science & Technology. 2012;**114**:1447-1459. DOI: 10.1002/ejlt.201200199

[59] Biswas A, Adhvaryu A, Gordon SH, Erhan SZ, Willett JL. Synthesis of diethylamine-functionalized soybean oil. Journal of Agricultural & Food Chemistry. 2005;**53**:9485-9490. DOI: 10.1021/jf050731o

[60] Bailosky LC, Bender LM, Bode D, Choudhery RA, Craun GP, Gardner KJ, et al. Synthesis of polyether polyols with epoxidized soy bean oil. Progress in Organic Coatings. 2013;**76**:1712-1719. DOI: 10.1016/j.porgcoat.2013.05.005

[61] Lu Y, Larock RC. New hybrid latexes from a soybean oil-based waterborne polyurethane and acrylics via emulsion polymerization. Biomacromolecules. 2007;**8**:3108-3114. DOI: 10.1021/bm700522z

[62] Zhong B, Shaw C, Rahim M, Massingill J. Novel coatings from soybean oil phosphate ester polyols. Journal of Coatings Technology. 2001;**73**:53-57. DOI: 10.1007/BF02730031

[63] Allauddin S, Somisetti V, Ravinder T, Rao B, Narayan R, Raju K. One-pot synthesis and physicochemical properties of high functionality soy polyols and their polyurethane-urea coatings. Industrial Crops and Products. 2016;**85**:361-371. DOI: 10.1016/j.indcrop.2015.12.087

[64] Liu W, Chen T, Xie T, Qiu R. Soybean oil-based thermosets with N-vinyl-2-pyrrolidone as crosslinking agent for hemp fiber composites. Composites Part A: Applied Science and Manufacturing. 2016;**82**:1-7. DOI: 10.1016/j.compositesa.2015.11.035

[65] Sharmin E, Zafar F, Akram D, Alam M, Ahmad S. Recent advances in vegetable oils based environment friendly coatings: A review. Industrial Crops and Products. 2015;**76**:215-229. DOI: 10.1016/j.indcrop.2015.06.022

[66] Lu Y, Larock RC. Soybean-oil-based waterborne polyurethane dispersions: Effects of polyol functionality and hard segment content on properties. Biomacromolecules. 2008;**9**:3332-3340. DOI: 10.1021/bm801030g

[67] Karadeniz K, Calikolu Y, Sen MY. A novel polyurethanes from epoxidized

soybean oil synthesized by ring opening with bifunctional compounds. Polymer Bulletin. 2017;**74**:2819-2839. DOI: 10.1007/s00289-016-1867-0

[68] Dahlke B, Hellbardt S, Paetow M, Zech WH. Polyhydroxy fatty acids and their derivatives from plant oils. Journal of the American Oil Chemists' Society. 1995;**72**:349-353. DOI: 10.1007/BF02541095

[69] Petrovic ZS, Guo A, Zhang W. Structure and properties of polyurethanes based on halogenated and nonhalogenated soy-polyols. Journal of Polymer Science, Part A: Polymer Chemistry. 2000;**38**:4062-4069. DOI: 10.1002/1099-0518(20001115)38:22<4062::AID-POLA60>3.0.CO;2-L

[70] Guo Y, Hardesty JH, Mannari VM, Massingill JL Jr. Hydrolysis of epoxidized soybean oil in the presence of phosphoric acid. Journal of the American Oil Chemists Society. 2007;**84**:929-935. DOI: 10.1007 / s11746-007-1126-5

[71] Sharma V, Kundu PP. Condensation polymers from natural oils. Progress in Polymer Science. 2008;**33**:1199-1215. DOI: 10.1016/j.progpolymsci.2008.07.004

[72] Petrović ZS, Zhang W, Zlatanić A, Lava CC, Ilavskyý M. Effect of OH/NCO molar ratio on properties of soy-based polyurethane networks. Journal of Polymers & the Environment. 2002;**10**:5-12. DOI: 10.1023/A:1021009821007

[73] Guo A, Zhang W, Petrovic ZS. Structure–property relationships in polyurethanes derived from soybean oil. Journal of Materials Science. 2006;**41**:4914-4920. DOI: 10.1007/s10853-006-0310-6

[74] Kolanthai E, Sarkar K, Meka SRK, Madras G, Chatterjee K. Copolyesters from soybean oil for use as resorbable biomaterials. ACS Sustainable Chemistry and Engineering. 2015;**3**:880-891. DOI: 10.1021/acssuschemeng.5b00001

[75] Tamami B, Sohn S, Wilkes G. Incorporation of carbon dioxide into soybean oil and subsequent preparation and studies of nonisocyanate polyurethane networks. Journal of Applied Polymer Science. 2004;**92**:883-891. DOI: 10.1002/app.20049

[76] Poussard L, Mariage J, Grignard B, Detrembleur C, Jerome C, Calberg C, et al. Non-Isocyanate polyurethanes from carbonated soybean oil using monomeric or oligomeric diamines to achieve thermosets or thermoplastics. Macromolecules. 2016;**49**:2162-2171. DOI: 10.1021/acs.macromol.5b02467

[77] Xia Y, Larock RC. Vegetable oil-based polymeric materials: Synthesis, properties, and applications. Green Chemistry. 2010;**12**:1893. DOI: 10.1039/c0gc00264j

[78] Scala JL, Wool RP. The effect of fatty acid composition on the acrylation kinetics of epoxidized triacylglycerols. Journal of the American Oil Chemists Society. 2002;**79**:59-63. DOI: 10.1007/s11746-002-0435-4

[79] Yan M, Lin F-Y, Cochran EW. Dynamics of hyperbranched polymers derived from acrylated epoxidized soybean oil. Polymer (United Kingdom). 2017;**125**:117-125. DOI: 10.1016/j.polymer.2017.07.088

[80] Liu W, Fei M-E, Ban Y, Jia A, Qiu R. Preparation and evaluation of green composites from microcrystalline cellulose and a soybean-oil derivative. Polymer. 2017:9, 541. DOI: 10.3390/polym9100541

[81] Jong L, Liu Z. Biobased composites from crosslinked soybean oil and thermoplastic polyurethane. Polymer Engineering and Science.

2017;**57**:275-282. DOI: 10.1002/pen.24410

[82] Obradovic J, Voutilainen M, Virtanen P, Lassila L, Fardim P. Cellulose fibre-reinforced biofoam for structural applications. Materials. 2017;**10**:619. DOI: 10.3390/ma10060619

[83] Chen C, Podolsky JH, Hernandez NB, Hohmann AD, Williams RC, Cochran EW. Preliminary investigation of bioadvantaged polymers as sustainable alternatives to petroleum-derived polymers for asphalt modification. Materials and Structures/Materiaux et Constructions. 2017;**50**:225. DOI: 10.1617/s11527-017-1097-4

[84] Dweib M, Hu B, Shenton Iii H, Wool R. Bio-based composite roof structure: Manufacturing and processing issues. Composite Structures. 2006;**74**:379-388

[85] Temmink R, Baghaei B, Skrifvars M. Development of biocomposites from denim waste and thermoset bio-resins for structural applications. Composites Part A: Applied Science and Manufacturing. 2018;**106**:59-69. DOI: 10.1016/j.compositesa.2017.12.011

[86] Zhan M, Wool RP. Biobased composite resins design for electronic materials. Journal of Applied Polymer Science. 2010;**118**:3274-3283. DOI: 10.1002/app.32633

[87] Liu W, Xie T, Qiu R. Improvement of properties for biobased composites from modified soybean oil and hemp fibers: Dual role of diisocyanate. Composites Part A: Applied Science and Manufacturing. 2016;**90**:278-285. DOI: 10.1016/j.compositesa.2016.07.018

[88] Liu W, Fei M-E, Ban Y, Jia A, Qiu R, Qiu J. Concurrent improvements in crosslinking degree and interfacial adhesion of hemp fibers reinforced acrylated epoxidized soybean oil composites. Composites Science and Technology. 2018;**160**:60-68. DOI: 10.1016/j.compscitech.2018.03.019

[89] Sousa AF, Ferreira S, Lopez A, Borges I, Pinto RJB, Silvestre AJD, et al. Thermosetting AESO-bacterial cellulose nanocomposite foams with tailored mechanical properties obtained by Pickering emulsion templating. Polymer (United Kingdom). 2017;**118**:127-134. DOI: 10.1016/j.polymer.2017.04.073

[90] Panda S, Mohanty P, Behera D, Kumar Bastia T. ESOA modified unsaturated polyester hybrid networks: A new perspective. Journal of Applied Polymer Science. 2016;**133**:44345. DOI: 10.1002/app.44345

[91] Trino ASM, Costa CSMF, Fonseca AC, Barata I, Julio E, Serra AC, et al. Novel composites from green unsaturated polyesters and fly ashes: Preparation and characterization. Reactive and Functional Polymers. 2016;**106**:24-31. DOI: 10.1016/j.reactfunctpolym.2016.07.004

[92] Wu Y, Li K. Replacement of styrene with acrylated epoxidized soybean oil in an unsaturated polyester resin from propylene glycol, isophthalic acid, and maleic anhydride. Journal of Applied Polymer Science. 2016;**133**:43052. DOI: 10.1002/app.43052

[93] Wu Y, Li K. Replacement of styrene with acrylated epoxidized soybean oil in an unsaturated polyester resin from propylene glycol and maleic anhydride. Journal of Applied Polymer Science. 2017;**134**:45056. DOI: 10.1002/app.45056

[94] Wu Y, Li K. Acrylated epoxidized soybean oil as a styrene replacement in a dicyclopentadiene-modified unsaturated polyester resin. Journal of Applied Polymer Science. 2018;**135**:46212. DOI: 10.1002/app.46212

[95] Liu W, Xie T, Qiu R. Biobased thermosets prepared from rigid isosorbide and flexible soybean oil

derivatives. ACS Sustainable Chemistry and Engineering. 2017;5:774-783. DOI: 10.1021/acssuschemeng.6b02117

[96] Yang Y, Shen M, Huang X, Zhang H, Shang S, Song J. Synthesis and performance of a thermosetting resin: Acrylated epoxidized soybean oil curing with a rosin-based acrylamide. Journal of Applied Polymer Science. 2017;**134**: 44545. DOI: 10.1002/app.44545

[97] Senoz E, Stanzione Iii JF, Reno KH, Wool RP, Miller MEN. Pyrolyzed chicken feather fibers for biobased composite reinforcement. Journal of Applied Polymer Science. 2013;**128**: 983-989. DOI: 10.1002/app.38163

[98] Li C, Xiao H, Wang X, Zhao T. Development of green waterborne UV-curable vegetable oil-based urethane acrylate pigment prints adhesive: Preparation and application. Journal of Cleaner Production. 2018;**180**:272-279. DOI: 10.1016/j.jclepro.2018.01.193

[99] Liu Z, Tisserat BH. Coating applications to natural fiber composites to improve their physical, surface and water absorption characters. Industrial Crops and Products. 2018;**112**:196-199. DOI: 10.1016/j.indcrop.2017.12.002

[100] Baturk E, Kahraman MV. Photocrosslinked biobased phase change material for thermal energy storage. Journal of Applied Polymer Science. 2016;**133**:1-8. DOI: 10.1002/app.43757

[101] Chen Z, Wu JF, Fernando S, Jagodzinski K. Soy-based, high biorenewable content UV curable coatings. Progress in Organic Coatings. 2011;**71**:98-109. DOI: 10.1016/j.porgcoat.2011.01.004

[102] Auclair N, Kaboorani A, Riedl B, Landry V. Acrylated betulin as a comonomer for bio-based coatings. Part II: Mechanical and optical properties. Industrial Crops and Products. 2016;**82**:118-126. DOI: 10.1016/j.indcrop.2015.11.081

[103] Dai J, Liu X, Ma S, Wang J, Shen X, You S, et al. Soybean oil-based UV-curable coatings strengthened by crosslink agent derived from itaconic acid together with 2-hydroxyethyl methacrylate phosphate. Progress in Organic Coatings. 2016;**97**:210-215. DOI: 10.1016/j.porgcoat.2016.04.014

[104] Dai J, Jiang Y, Liu X, Wang J, Zhu J. Synthesis of eugenol-based multifunctional monomers via a thiol-ene reaction and preparation of UV curable resins together with soybean oil derivatives. RSC Advances. 2016;**6**:17857-17866. DOI: 10.1039/c6ra01420h

[105] Lee S, Park MS, Shin J, Kim Y-W. Effect of the individual and combined use of cardanol-based plasticizers and epoxidized soybean oil on the properties of PVC. Polymer Degradation and Stability. 2018;**147**:1-11. DOI: 10.1016/j.polymdegradstab. 2017.11.002

[106] Kandula S, Stolp L, Grass M, Woldt B, Kodali D. Functionalization of soy fatty acid alkyl esters as bioplasticizers. Journal of Vinyl and Additive Technology. 2017;**23**:93-105. DOI: 10.1002/vnl.21486

[107] Panaitescu DM, Nicolae CA, Frone AN, Chiulan I, Stanescu PO, Draghici C, et al. Plasticized poly(3-hydroxybutyrate) with improved melt processing and balanced properties. Journal of Applied Polymer Science. 2017;**134**:44810. DOI: 10.1002/app.44810

[108] Darie-Ni RN, Vasile C, Irimia A, Lipa R, Rp M. Evaluation of some eco-friendly plasticizers for PLA films processing. Journal of Applied Polymer Science. 2016;**133**:43223. DOI: 10.1002/app.43223

[109] Xing C, Matuana LM. Epoxidized soybean oil-plasticized poly(lactic acid) films performance as impacted

by storage. Journal of Applied Polymer Science. 2016;**133**:43201. DOI: 10.1002/app.43201

[110] Li C, Sung J, Sun XS. Network from Dihydrocoumarin via solvent-free metal-mediated pathway: A potential structure for substantial toughness improvement of epoxidized plant oil materials. ACS Sustainable Chemistry and Engineering. 2016;**4**:1231-1239. DOI: 10.1021/acssuschemeng.5b01283

[111] Tsujimoto T, Ohta E, Uyama H. Plant oil-based shape memory polymer using acrylic monolith. Express Polymer Letters. 2015;**9**:757-763. DOI: 10.3144/expresspolymlett.2015.71

[112] Tsujimoto T, Uyama H. Full biobased polymeric material from plant oil and poly(lactic acid) with a shape memory property. ACS Sustainable Chemistry and Engineering. 2014;**2**:2057-2062. DOI: 10.1021/sc500310s

[113] Miao S, Liu K, Wang P, Su Z, Zhang S. Preparation and characterization of epoxidized soybean oil-based paper composite as potential water-resistant materials. Journal of Applied Polymer Science. 2015;**132**. DOI: 10.1002/app.41575

[114] Huang X, Wang A, Xu X, Liu H, Shang S. Enhancement of hydrophobic properties of cellulose fibers via grafting with polymeric epoxidized soybean oil. ACS Sustainable Chemistry and Engineering. 2017;**5**:1619-1627. DOI: 10.1021/acssuschemeng.6b02359

Chapter 9

Peptides and Microorganisms Isolated from Soybean Sources with Antimicrobial Activity

Rosalva Mora-Escobedo, María Del Carmen Robles-Ramírez, Alma Delia Román-Gutiérrez, Javier Castro-Rosas, Ciro Baruchs Muñoz-Llandes and Fabiola Araceli Guzmán-Ortiz

Abstract

Soybean has a high biological value because it is a potential nutraceutical that benefits human health. Isolated peptides of soybean have been associated with activities such as anticancer, antioxidants, antiobesity, antithrombotics, hypocholesterolemic, antidiabetic, immunomodulatory and antimicrobial, and this last activity is also obtained from microorganisms isolated from soybean subjected to processes such as fermentation, which can act directly against pathogens that are resistant to antibiotics or participate in the synthesis of new peptides with antimicrobial activity, so they could be used as preservatives in food as an alternative in the prevention of diseases. Strains of *Bacillus subtilis* isolated from soybean are mainly those that have the ability to inhibit the growth and proliferation of pathogens; some fungi such as *Rhizopus microspores* and *Aspergillus oryzae* have also had an inhibitory effect. This chapter describes the potential of microorganisms and peptides obtained from different sources of soybean against pathogenic microorganisms responsible for foodborne diseases.

Keywords: soybean, antimicrobial, peptides, microorganisms, fermentation

1. Introduction

Soy is a vegetable source of high protein content that simultaneously has several beneficial effects in human health. Its composition includes nutritional compounds of high biological value as lipids, vitamins, minerals, sugars, isoflavones, flavonoids, saponins, and peptides. Its nutritional content has proven its antihypertensive, anticholesterolemic, antioxidant, and anticancer activity [1]. Soy fermentation has mainly been conducted to preserve foods. However, it has been proven that this process is generated by certain microorganisms, expresses diverse nutritional compounds of certain biological value, and simultaneously lowers the concentration of antinutritional components as proteases, phytic acid, urease, and oxalic acids [2]. During this process, starter microorganisms turn complex organic compounds into simple compounds, improving their functionality and availability in the food matrix [3].

Different foods are derived from soy fermentation, in which lactic acid bacteria are used as starter microorganisms, mostly bacteria belonging to the genus *Bacillus*. These bacteria are responsible for the expression of certain antimicrobial components and provide the fermented products with certain organoleptic characteristics. Peptides, a product of protein hydrolysis, are one of the components expressed by starter microorganisms. They have been proven to exert a biological activity, largely due to the antagonistic and inhibitive effect of pathogenic microorganisms that cause gastrointestinal diseases and are transmitted by food. It has been demonstrated that short-chain peptides have a more efficient inhibitive action than long-chain ones [4].

On the other hand, it has been reported that some antimicrobial lipopeptides can interact with the cell's membrane, forming pores and leading to cell lysis [5].

This chapter consolidates relevant information on microorganisms and peptides isolated from soy-based foods that are known to have a microbial and/or antagonistic action in the presence of pathogenic bacteria. Reports are listed stating that, after fermentation, microorganism can express antimicrobial compounds that could be analyzed to be substituted by antibiotics.

2. Fermentation

Fermentation is a feasible way to produce peptides when compared against the use of enzymes in protein hydrolysis. Proteolytic enzymes produced by microorganisms involved in fermentation release peptides and free amino acids [6, 7]. For instance, the β-galactosidases in lactic acid bacteria in fermentation can hydrolyze oligosaccharides present in soy, reducing the unpleasant taste and flatulence [8–10]. Oligopeptides, dipeptides, and tripeptides are created as a result of protein hydrolysis by this type of proteolytic enzymes in soy milk fermentation [11]. The creation of peptides also depends on the soybean variety, the type of inoculation microorganism, and even the strain. Sanjukta et al. [7] found that the efficiency of hydrolyzed protein and free amino acids in fermented soybean with *Bacillus subtilis* MTCC5480 was higher than with *Bacillus subtilis* MTCC1747. The activity and generation of microorganisms during fermentation can vary depending on the substrate to produce a determined food. Among the possible sources of fermented-soybean products created from *Bacilli* are natto, kinema, and chungkookjang. Some other products as sufu, tempeh, douchi, miso, and combinations as doenjang are obtained from fungi. Other products derived from soybean as doubanjiang, meju, sokseongjang, cheonggukjang, kanjang, thua nao, hawaijar, and tungrymbai also create microorganisms with certain biological functions [6, 12, 13].

3. Fermented soybean products

3.1 Buckwheat sokseongjang

Buckwheat sokseongjang, a traditional Korean food, is an aged paste made from fermented soybeans. This fermentation takes place with *Bacillus subtilis* HJ18-4 as inoculum at 35°C for 36 h [14]. During the fermentation of buckwheat sokseongjang, there is a decrease in sugar content while the opposite occurs for protease and amylase activities [15]. Due to the type of microorganism, the soybean food shows a high antimicrobial spectrum. *Lactobacillus plantarum* JSA22 has also been isolated from buckwheat sokseongjang with probiotic properties that are able to inhibit infection by *S. typhimurium* in intestinal epithelial cells [16].

3.2 Cheonggukjang

This food is also produced from fermented soybeans and, as buckwheat sokseongjang, it is traditional in Korea [17]. Fermentation of cheonggukjang is completed at 40–43°C for 48–96 h [18], using natural microflora as *Bacillus subtilis* [19, 20]. During this process, the hydrolysis of compounds increases, leading to the conversion of glycosylated flavonoids into aglyconated forms. Additionally, several proteins are degraded into small peptides and amino acids [21, 22]. Cheonggukjang contains many enzymes, microorganisms, and bioactive compounds considered a source of proteins, hydrolyzed peptides, and lipids [19, 20, 23].

3.3 Miso

Miso is a soybean paste obtained by fermentation with lactic acid, yeast, and tane-koji (starter). It maintains some of the texture of the beans; however, the final product is a paste. *Aspergillus oryzae* strains are inoculated, allowing the fungus to grow and cover the beans; the process can take months and the same miso can be used as inoculum for new fermentations [24]. The presence of microorganisms as *Lactococcus* sp. GM005, which are capable of producing peptides with antimicrobial activity has been reported in miso fermentation [25].

3.4 Solid fermented soy foods

3.4.1 Douchi

Within the solid fermentation of soy products is douchi, a product of black soybeans that is traditional in China. Several types of douchi can be identified according to the type of microorganisms used in the fermentation (filamentous fungi or bacteria). The soybeans are washed and soaked for 3–4 h. Then, they are steamed for 50 min approximately and cooled at 30°C. The beans are inoculated with *Aspergillus aegyptiacus* [26], although *Mucor*, *Rhizopus*, and bacteria can also be used [27]. After inoculation, the beans are maintained at 30°C for 3–4 days to make koji. Afterwards, they are washed with water and mixed with 16% salt, water, ginger, and a mix of powdered spices. During maturation, the fungal enzymatic activity promotes an increase in amino nitrogen levels and organic acid as well as a reduction in the concentration of isoflavones [24, 26]. Some studies have reported that douchi inhibits the angiotensin-converting enzyme and α-glucosidase and has an antioxidant activity [28, 29].

3.4.2 Natto and kinema

Natto and kinema are soy products with similar elaboration processes. To make natto, soybeans are soaked in water at 21–23°C for 20 h and are boiled at 121°C for 40 min. They are then cooled at 50°C for inoculation with a pure culture of *Bacillus subtilis* natto spores that germinate at 50°C [30]. Similarly, to produce kinema, soybeans are soaked overnight at 25°C, cooked, and ground [31]. The fermentation of the product occurs naturally: during the process, bacterial spores able to survive cooking are generated; therefore, there is no need for starter microorganisms. Due to the type of microbiota generated during the fermentation of kinema (species of *Bacillus*, *Bacillus subtilis*, *Bacillus licheniformis*, *Bacillus cereus*, *Bacillus circulans*, *Bacillus thuringiensis*, and *Bacillus sphaericus* [32] some antimicrobial activity is likely to exist, mainly because of the presence of *Bacillus subtilis*, which has been reported to have an antimicrobial effect [33].

3.4.3 Tempeh

Tempeh is another solid fermented soy product. Its preparation starts with hull removal; then, the beans are soaked in water for 17 h, approximately. They are cooked in water for 30–40 min; the water is drained and the beans are cooled at room temperature for inoculation with *Rhizopus oligosporus* and incubated at 32°C for 26 h [34]. The fungal metabolism created during the fermentation causes changes in the composition of tempeh due to the increased protein degradation generating molecules of low molecular weight [35]. Some studies have reported that peptides, as bacteriocins, are generated from microorganisms isolated from tempeh and have an antibiotic effect [36].

4. Antimicrobial peptide action

The process of soy fermentation creates secondary metabolites, as antibiotics and peptides that have some biological activity beneficial to human health. Antimicrobial ability is one of the reported activities that secondary metabolites exhibit. It may be due to the peptides produced by protein hydrolysis present in the food or starter culture. It has been reported that *Bacillus subtilis* can produce a wide range of antimicrobial compounds with an antagonistic effect against bacteria and fungi [12, 14, 37, 38]. It is known that *Bacillus subtilis* is able to create bioactive peptides. Some pathogenic microorganisms susceptible to these peptides are: *Campylobacter* spp, *Clostridium botulinum*, *Listeria monocytogenes*, *Salmonella typhimurium*, *Staphylococcus aureus*, *Pseudomonas aeruginosa*, *Botrytis cinerea*, *Fusarium moniliforme*, *Micrococcus luteus*, and *S. typhimurium* [39]. The action mechanism of the antimicrobial activity can be different between peptides. Their hydrophobic characteristics allow them to interact with the lipid layer of the cell membrane. Those peptides in close contact with the bacterial cell must cross the capsular polysaccharide to interact with the outer membrane. Once the peptides have connected with the plasma membrane, they can interact with the lipid bilayer [40]. Through the interaction with the bacterial membrane, peptides cover the cell surface and lipids are aggregated. In consequence, peptides interact with the membrane and destabilize it, leading to pore formation and disruption

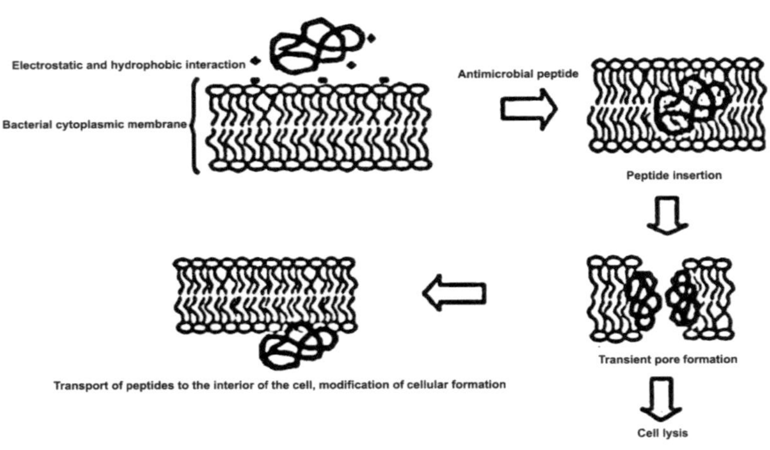

Figure 1.
Action mechanism of antimicrobial peptides (modified from Ruiz et al. [43]).

of the membrane, leading to cell lysis (**Figure 1**) [41–43]. In addition, peptides can produce indentations in the cell surface and induce micellization, disrupting the bacterial membrane [44]. On the other hand, an extensive presence of bubbles in bacterial cells promotes the release of intracellular material and, in consequence, antimicrobial peptides [45]. When peptides penetrate into the cells, they create clusters and aggregates, which promotes ion leakage from the cell [42]. The action mechanisms can depend on a number of physicochemical factors of the peptide's structure, charge, chain length, amino acid composition, and amino acid sequence.

5. Microorganisms

The fermentation quality of soy foods is determined by the type of bacteria generated. Lactic acid bacteria are especially interesting because of their function. They provide flavor and protect against contamination by pathogenic microorganisms [24]. These bacteria can produce bacteriocins in fermented-soybean products that contribute to control the growth of microorganisms as *Listeria* sp. and *Klebsiella* spp. Strain *Bacillus subtilis* is able to produce antibiotics and peptides as bacteriocins and lipopeptides with antimicrobial activity. This genus is a strong producer of hydrolytic enzymes that allow for the digestibility of fermented-soybean products [46]. The most common contamination in fermented products is caused by *Bacillus cereus*; this microorganism can produce enterotoxins on cooked soybeans, i.e., in kinema. However, only small amounts of enterotoxins were formed in the presence of a competitive dominance of *Bacillus subtilis* [47].

5.1 Lactic acid bacteria

Lactic acid bacteria are considered probiotic microorganisms thanks to the diverse beneficial effects they produce on human health. Still, they have also been proven to have antimicrobial activity against pathogenic bacteria [48]. Recent studies on probiotics have reported that fermented-soybean products have strong antibactericidal effects against pathogens transmitted by food [49]. These bacteria show inhibitive effects on many pathogenic organisms both *in vivo* and *in vitro*, including *Salmonella*, *Shigella*, *Clostridium*, *Bacillus cereus*, *Staphylococcus aureus*, *Candida albicans*, *Listeria monocytogenes*, *Escherichia coli*, and *Campylobacter jejuni* [50]. El-Sayed et al. [51] analyzed the antimicrobial activity of probiotic lactic acid bacteria during the fermentation process of soymilk. They found those microorganisms have an antagonistic effect against pathogenic agents as *E. coli* and *S. aureus*. Additionally, the fermentation process generates secondary metabolites as organic acids, hydrogen peroxide, and certain bacteriocins, which enhance the inhibitory capacity of pathogenic microorganisms. Bacteriocins are ribosomally synthesized antimicrobial peptides with inhibitory activity towards a wide range of pathogens [52]. Lactic acid bacteria isolated from *Tungtap*, a fermented species of Puntius, and *Tungrymbai*, a fermented-soy product, can produce bacteriocins. The bacterocins of *lactobacillus*, *Pediococcus* and *Enterococcus* possess better inhibitory properties against pathogens than antibiotics. Bacteriocins have an inhibitory effect against β-lactamase greater than antibiotics, which is desired because this enzyme provides resistance to bacteria to be inhibited [53]. *Streptococcus pyogenes*, *Enterococcus faecalis*, *Escherichia coli*, *Klebsiella pneumoniae*, and *Bacillus cereus* are the main microorganisms inhibited by this type of bacteriocins produced from lactic acid bacteria.

5.2 *Bacillus subtilis* SN7

Bacillus subtilis strain SN7 is isolated from meju, a fermented-soy product from Korea. Strain SN7 has an antimicrobial activity against pathogens as *Bacillus cereus* and is able to produce bacteriocins. This strain poses no risk to human health and is therefore widely recommended. It has been used in the fermentation of cheonggukjang. *Bacillus subtilis* SN7 efficiently inhibits the growth of *Bacillus cereus* vegetative cells and the inactivated germination of *Bacillus cereus* spores [54]. This is because when 4.2–4.3 log spores/mL of *B. cereus* ATCC 14579 and 6.2–6.3 log colony forming units (CFU) /mL of *B. subtilis* SN7, were inoculated at 37° C for 24 h in tryptic soy broth, *B. cereus* is not detected at 12 h of culture. The antimicrobial effect of *Bacillus subtilis* SN7 is mainly due to its capacity in the production of bacteriocins, and these act directly on the cell wall of the bacteria causing death and lysis of the cell [55]. This suggests that *Bacillus subtilis* SN7 has a great potential as control agent of *Bacillus cereus* and starter culture in Cheonggukjang. It has a wide-spectrum antagonistic activity against *Bacillus cereus*, *Bacillus subtilis*, *Bacillus licheniformis*, *Staphylococcus aureus*, *E. coli* O157: H7, and *Micrococcus luteus*. However, after extended incubation periods, the antimicrobial activity can be inhibited since the antimicrobial compound is a protein and thus it is destroyed by proteolytic enzymes produced by the same *Bacillus subtilis* SN7. Most of the *Bacillus* strains produce proteolytic enzymes at the end of the growth for spore extrusion through autocellular lysis [56]. The antimicrobial capacity produced by bacteriocin of *Bacillus subtilis* SN7 it can be inactivated by proteolytic enzymes but not by α-amylase or lipase. Mejucin, the bacteriocin produced by *Bacillus subtilis* SN7, has a sequence LGPQLNKGCATCS IGAACLVDGPIPDEIAG. The pH conditions do not affect the antimicrobial activity since Mejucin tolerates pH 3–9 and is stable between 4 and 37°C [54]. Mejucin modifies the cell membrane in such a way that its regeneration is not possible, and it also causes the leakage of intracellular components of *Bacillus cereus* [13].

5.3 *Bacillus* sp. LM7

Bacillus sp. LM7 is a microorganism isolated from Chungkookjang. During the stationary growth phase of *Bacillus* sp. LM7 and soy fermentation, the microorganisms releases an antimicrobial substance called anti-LM7, which has been reported to be stable at pH 4–9 and 80°C for 30 min. The reported antimicrobial activity mainly affects gram-positive bacteria, including *Bacillus cereus* and *Listeria monocytogenes*, and inhibits certain types of fungal strains. The antimicrobial effect is associated to the generation of lipopeptides from two families, bacillomycin D and surfactin analogues [57], so anti-LM7 was considered a lipopeptide. Bacillomycin D is composed of L-Asn, D-Tyr, D-Asn, L-Pro, L-Glu, D-Ser, and L-Thr [58, 59]. Surfactin consists of a hydrophobic fatty acid and seven amino acids, including L-Asp, 2 L-Leu, L-Glu, L-Val, and 2 D-Leu. However, its efficiency is negatively affected by enzymes (trypsins, lipases, pepsins, or proteinase k). Chymotrypsin cleaves the amide bond on the carboxyl side of tyrosine, tryptophan, phenylalanine, and leucine. The protease hydrolyzes peptide amide bonds on the carboxyl side of glutamic and aspartic acids. Therefore, chymotrypsin and protease can inhibit the anti-LM7 activity by recognizing and hydrolyzing an amino acid in the peptide loop. Chymotrypsin can hydrolyze peptide bonds on the carboxyl side of Tyr, Trp, Phe and Leu and the protease Glu and ASP. Anti-LM7 isolated from chungkookjang has a potential use in environmental, pharmaceutical, and food-processing industries, and this can be used in dairy products and in the production of vegetables and meats as a preservative to increase the food's shelf life [57]. It could also be an antibiotic source, considering the conditions under which its activity can be affected.

5.4 *Bacillus subtilis* SC-8

Bacillus subtilis SC-8 is a gram-positive bacterium that has been isolated from fermented soy. The microorganism inhibits the growth of *Bacillus cereus* species as *Bacillus anthracis*, *Bacillus mycoides*, *Bacillus pseudomycoides*, *Bacillus thuringiensis*, and *Bacillus weihenstephanensis* [60, 61]. Genes were also found on chromosome SC-8 of *Bacillus subtilis* (BSSC8_21740 to BSSC8_21920), these genes can code new antimicrobial peptides against *B. cereus*. [62]. As *Bacillus* sp. LM7, *Bacillus subtilis* SC-8 can express surfactin, fengycin, and iturin. Fengycin is an antifungal that inhibits filamentous fungi, yeasts, and bacteria. Its structure is given by: Glu-D-Orn-D-Tyr-DaThr-L-Glu-D-Al/Val-L-Pro-L-Gln-L-Tyr-L-Ile. The iturin family includes bacillomycin, iturin, and mycosubtilin; they are cyclic lipoheptapeptides bound by a β-amino acid residue. The members of this family have a strong antibiotic property. Its structure is given by iturin A, FA-β-NH2-L-Asn-D-Tyr-D-Asn-LGln-L-Pro-D-Asn-L-Ser and iturin C, FA-β-NH2-L-Asp-D-Tyr-D-Asn-LGln-L-Pro-D-Asn-L-Ser [5].

5.5 *Bacillus subtilis* NT-6

Bacillus subtilis NT-6 has been isolated from natto; it is able to release AMPNT-6, an amphipathic peptide with a hydrophobic fatty acid section and a hydrophilic peptide section. The molecular structures of these lipopeptides are cyclic because the C-terminal peptide residues connect with the β-hydroxy fatty acid through the hydroxyl group of the peptide residue directly with a β-amino acid [63]. Peptide AMPNT-6 has antimicrobial characteristics within a wide pH range (2–12). It is thermally stable, even at 100°C during 20 min, and has excellent solubility in both oil and water. It is efficient not only against gram-positive and gram-negative bacteria but also against fungi [64–66]. Xu et al. [67] reported a study on AMPNT-6 against *Vibrio parahaemolyticus*. The antimicrobial efficiency of the peptide depends on time and concentration. AMPNT-6 has a high inhibitory effect on *V. parahaemolyticus* at a minimal inhibitory concentration (MIC) (1.25 mg/mL). The action mechanism is the result of the destruction of the cell wall, forming pores in the cell membranes. Peptide AMPNT-6 has also been studied to determine its capacity to reduce adhesion and alter *Shewanella putrefaciens* preformed biofilms in two different contact surfaces (shell of shrimp, stainless steel blade). At minimal inhibitory concentration of 3 mg/mL, it was able to eliminate the formation of biofilms and prevent bacteria from forming them again in a model using a 96-well polystyrene microplate. It has also been reported that AMPNT-6, decreases the amount of extracellular polymeric substances secreted by bacteria, which is beneficial, since these substances are the main component of the biofilm responsible for adhesion [68, 69] and can strengthen the interactions between bacterial cells and participate in the formation of bacterial colonies on the contact surface [70]. At MIC of 2 mg/mL, it prevents bacteria from adhering to the surface of the microplate to form biofilms in a 3 h period while the elimination of formed biofilms takes place at 24 h [71]. *Staphylococcus aureus* and enterotoxin B, have also been controlled by the AMPNT-6 peptide, this may be due to the fact that they have been reported to contain three main lipopeptides such as surfactin, fengycin and iturin, capable of producing pores in the membranes that do not generate strains resistant to drugs [67, 72–74]. The antimicrobial activity of the AMPNT-6 peptide is not affected by the conditions used in meat products (temperature 25–37°C, sodium chloride concentration 7–8%, pH 7.4–8.4, and sodium metabisulfite concentration 0.2–0.4%); therefore, it can be viable for application in meat products [66].

5.6 *Bacillus subtilis* HJ18-4

Bacillus subtilis HJ18-4 is a microorganism that has been isolated from buckwheat sokseongjang. It has an antimicrobial activity against *Bacillus cereus* and other pathogens [75]. To demonstrate the antimicrobial effect of *B. subtilis* against *B. cereus*, these microorganisms have been inoculated in Luria-Bertani broth at 30° C for 24 h. *B. subtilis* has been inoculated at different concentrations (0.125, 0.25, 0.5 and 1%) and *B. cereus* at 0.5%. The highest inhibitory effect that *B. subtilis* showed against *B. cereus* was 0.5 and 1%, the survival of *B. cereus* decreases by 6.87–5.65 log CFU/mL, respectively [14]. In addition, gene expression has been used to prove the efficacy of the microorganism's antimicrobial capacity. Eom et al. [14] reported a decrease in the expression of genes related to toxin of *Bacillus cereus* as groEL, nheA, nheC, and entFM with *Bacillus subtilis* HJ18-4, proving its antimicrobial capacity.

5.7 *Bacillus subtilis* SCK-2

Among the bacteria isolated from the traditional Korean paste of fermented soy Kyeopjang, *Bacillus subtilis* SCK-2 shows antimicrobial activity against *Bacillus cereus* [76]. Peptide AMPC IC-1 has been identified from *Bacillus subtilis* SCK-2. It is a 33-residue thermostable peptide of 13 amino acids (Cys, Asn or Asp, Gln or Glu, Ser, Ala, Pro, Gly, Arg, Thr, Val, Ile, Leu and Lys) and its molecular weight is 3.4–3.6 kDa [77]. It has antimicrobial properties against species from the *Bacillus cereus* group. In addition, AMPC IC-1 inhibits the growth of *Bacillus cereus* KCTC 3624, KCTC 3674, and KCTC 3711 at a concentration of 50 μg/mL for 24 h [78]. However, it is not efficient for other pathogenic microorganisms as *Listeria monocytogenes*, *Salmonella enterica*, *Salmonella enteritidis*, *Staphylococcus aureus*, and *Escherichia coli* O157:H7 [77]. The antimicrobial activity of AMPC IC-1 is more stable under neutral and alkaline conditions [78]. The interaction between this peptide and proteases and proteinase K reduce antimicrobial activity [12]. The antimicrobial action mechanism takes place through the permeation of the cell membrane [76].

5.8 *Enterococcus faecium*

Enterococcus faecium is predominant in fermented foods, where it plays a key role thanks to its contribution to maturation and aroma development [79]. It is able to produce bacteriocins and has an inhibitory effect against pathogenic bacteria as *Listeria* sp., *Staphylococcus aureus*, *Vibrio cholerae*, *Clostridium* sp., *Bacillus* sp., and *Helicobacter pylori* [80–86]. Two bacteriocins (1 and 2) were obtained in tempeh from *E. faecium* LMG 19827 and *E. faecium* LMG 1982, respectively. Bacteriocins 1 and 2 have a molecular weight of 3.4 and 5.4 kDa. Their antimicrobial activity is mainly against *L. monocytogenes* and gram-positive indicators as *Enterococcus faecalis*, *E. faecium*, *Carnobacterium divergens*, *C. piscicola*, *Lactobacillus brevis*, *L. pentosus*, and *Paralactobacillus selangorensis*. The antimicrobial activity of bacteriocins is more stable after thermal treatment, except at alkaline pH values [36].

Enterococcus faecium has been isolated from chungkookjang. The strain is resistant against simulated gastrointestinal diseases. Strains S2C10 and S2C11 inhibit the viability of *Listeria monocytogenes* ATCC 19113. *L. plantarum* KCTC 1048, *Lactococcus lactis* KCTC 1913, *L. lactis* KCTC 3769, *Leuconostoc lactis* ATCC 19256, and *Pediococcus acidilactici* KCTC 1627. The ability is probably due to bacteriocin production. This antimicrobial effect is not altered at pH 2–8; however, at pH 10, the activity is reduced and often affected by enzymes as lipase [87].

5.9 *Lactococcus* sp. GM005

Another microorganism isolated from a fermented-soybean product (miso paste) is *Lactococcus* sp. GM005, which can produce 9.6-kDa bacteriocins against *L. sakei* (JCM1157 [25, 88]. This bacteriocin has been called GM005 because of the strain it is originated in. It contains a high proportion of hydrophobic amino acid residues and lanthionine [89]. Bacteriocin GM005 is sensitive to trypsin, has shown antibacterial activity against a producer of nisin that is immune to it, stable to heat and neutral pH [25]. Colony forming units (CFU) per mL of *L. sakei* JCM1157T have shown a decrease in their growth when treated with GM005 [25].

5.10 *Bacillus subtilis* E20

Cheng et al. [33] identified three peptides with antimicrobial activity. The peptides were isolated in a protein fraction from a solid-state fermentation of soy flour with *Bacillus subtilis* E20. These peptides showed an activity against *Vibrio alginolyticus* and *V. parahaemolyticus.* They have been identified as LSKKHEAALKAFTDAQKQ (2.01 kDa), LRFRAPAPVLRRIAKR (1.96 kDa), and HTSKALLDMLKRLGK (1.71 kDa). This last peptide has shown a more efficient antimicrobial activity. The minimum inhibitory concentration reported is 72 mM [90]. The growth inhibition of *Vibrio* can be by the interaction with the bacterial membranes that cause an increase in permeability [41, 91].

5.11 *Bacillus natto* TK-1

Lipopeptides can be obtained from *Bacillus natto* TK-1 in natto. The ability of *Bacillus natto* TK-1 to express components with antibacterial and antifungal activities has been proven [92, 93]. The effect is due to the production of lipopeptides as fengycin, which affects the surface of the cell membrane, inhibiting the development of bacteria as *E. coli*, *Salmonella typhimurium*, and *Staphylococcus aureus* and fungi as *Botrytis cinerea* and *Fusarium moniliforme* [91, 93–94]. However, the efficacy of inhibition against these microorganisms and other pathogens depends on the method of inoculation and the concentration of the antimicrobial [92].

6. Other peptides

Peptides isolated directly from soybeans have also proven to have antimicrobial activity. Peptides PGTAVFK and IKAFKEATKVDKVVVLWTA are protein sources from soy evaluated against *Listeria monocytogenes* and *Pseudomonas aeruginosa*. The long-chain peptide (IKAFKEATKVDKVVVLWTA) is more effective against both microorganisms at an inhibitory concentration of 37.2 μM. Still, peptide PGTAVFK does not significantly affect the development and proliferation of *P. aeruginosa* but inhibits *L. monocytogenes* at a concentration higher than 625 μM [95, 96]. PGTAVFK also has an antimicrobial activity against *E. coli* and *S. aureus* at a 31 μM concentration [97].

Peptides can also be produced by gastrointestinal digestion. Digestive enzymes hydrolyze proteins, producing peptides of different sizes and free amino acids. In *in vitro* systems, pepsin is used to act at stomach level and randomly hydrolyzes peptide bonds to produce relatively large peptides, a mix of pancreatic acid and pancreatins. The pancreatins are a mix of different peptidases as trypsin, α-chymotrypsin, elastase, and carboxypeptidases. All the enzymes used, with the exception of trypsin, hydrolyze peptide bonds, producing peptides that differ in amino acid sequence

[98]. Peptides (613 and 4932 Da) generated by acylated protein of soy seed and soymilk and subject to an *in vitro* digestion process have proven to have biological effects. Their size is slightly larger when compared to peptides produced by animal digestion. Hydrolysis can be limited due to the presence of some antinutritional factors as trypsin, lectins, and oligosaccharides inhibitors and structural barriers (cell wall), which are scarcely digestible and inhibit diffusion of digestive enzymes [24, 99].

The analysis of the data base and algorithms of peptide sequences predicted the generation of 11 peptides obtained from soy-seed digestion, 17 from soymilk protein, and 8 from antimicrobial soymilk protein precipitate. The generation of nine of those peptides is the consequence of glycine and β-conglycinin digestion. The sequences of the antimicrobial peptides reported were: IIIAQGKGALGV, SGGIKLPTDIISKISPLPVLKEI, SGGIKLPTDIISKISPLPV, and MIIIAQGKGALGV, IIVVQGKGAIG [97].

7. Conclusion

Soy is an important source of bioactive compounds and proteins. Some bioactive peptides obtained from soy and its fermented and non-fermented products show different biological actions as the antimicrobial activity. The fermentation process allows for the generation of microorganisms, mainly *Bacillus subtilis*, able to inhibit the growth of pathogenic bacteria. These microorganisms can participate in the production of new peptides with antagonistic activity towards pathogens, resulting in cellular lysis. Peptides can also be produced by fermentation, starter cultures, or hydrolysis of soy proteins. These products can be incorporated as additives in the development of functional foods and pharmaceutical products for preservation and prevention of health risks.

Author details

Rosalva Mora-Escobedo[1], María Del Carmen Robles-Ramírez[1], Alma Delia Román-Gutiérrez[2], Javier Castro-Rosas[2], Ciro Baruchs Muñoz-Llandes[2] and Fabiola Araceli Guzmán-Ortiz[3]*

1 Departamento de Ingeniería Bioquímica, Escuela Nacional de Ciencias Biológicas, Instituto Politécnico Nacional, Unidad Adolfo López Mateos, Ciudad de México, Mexico

2 Área Académica de Química (AAQ), Universidad Autónoma del Estado de Hidalgo, Ciudad del Conocimiento, Mineral de la Reforma, Hidalgo, CP, México

3 CONACYT—Universidad Autónoma del Estado de Hidalgo, Ciudad del Conocimiento, Mineral de la Reforma, Hidalgo, CP, México

*Address all correspondence to: fabiguzman01@yahoo.com.mx, aroman@uaeh.edu.mx

References

[1] Kim SL, Berhow MA, Kim JT, Chi HY, Lee SJ, Chung IM. Evaluation of soyasaponin, isoflavone, protein, lipid, and free sugar accumulation in developing soybean seeds. Journal of Agricultural and Food Chemistry. 2006;**54**(26):10003-10010

[2] Egounlety M, Aworh OC. Effect of soaking, dehulling, cooking and fermentation with *Rhizopus oligosporus* on the oligosaccharides, trypsin inhibitor, phytic acid and tannins of soybean (*Glycine max* Merr.), cowpea (*Vigna unguiculata* L. Walp) and groundbean (*Macrotyloma geocarpa* Harms). Journal of Food Engineering. 2003;**56**(2-3):249-254

[3] Cho YS, Kim SK, Ahn CB, Je JY. Preparation, characterization, and antioxidant properties of gallic acid-grafted-chitosans. Carbohydrate Polymers. 2011;**83**(4):1617-1622

[4] Xiang N, Lyu Y, Zhu X, Bhunia AK, Narsimhan G. Effect of physicochemical properties of peptides from soy protein on their antimicrobial activity. Peptides. 2017;**94**:10-18

[5] Ariza Y. Determinación de metabolitos secundarios a partir de Bacillus subtilis efecto biocontrolador sobre Fusarium sp. Nova. 2012;**10**(18):135-250

[6] Rai AK, Jeyaram K. Health benefits of functional proteins in fermented foods. Health Benefits of Fermented Foods and Beverages. 2015;**44**:455-474

[7] Sanjukta S, Rai AK, Muhammed A, Jeyaram K, Talukdar NC. Enhancement of antioxidant properties of two soybean varieties of Sikkim Himalayan region by proteolytic Bacillus subtilis fermentation. Journal of Functional Foods. 2015;**14**:650-658

[8] Dhananjay S, Kulkarni S, Kapanoor KG, Naganagouda VK, Veerappa HM. Reduction of flatus-inducing factors in soymilk by immobilized-galactosidase. Biotechnology Applied Biochemistry. 2006;**45**:51-57

[9] Hati S, Vij S, Mandal S, Malik RK, Kumari V, Khetra Y. α-Galactosidase activity and oligosaccharides utilization by lactobacilli during fermentation of soy Milk. Journal of Food Processing and Preservation. 2014;**38**(3):1065-1071

[10] Scalabrini P, Rossi M, Spettoli P, Matteuzzi D. Characterization of bifidobacterium strains for use in soymilk fermentation. International Journal of Food Microbiology. 1998;**39**(3):213-219

[11] Agyei D, Danquah MK. Industrial-scale manufacturing of pharmaceutical-grade bioactive peptides. Biotechnology Advances. 2011;**29**(3):272-277

[12] Sanjukta S, Rai AK. Production of bioactive peptides during soybean fermentation and their potential health benefits. Trends in Food Science & Technology. 2016;**50**:1-10

[13] Lee SG, Chang HC. Purification and characterization of mejucin, a new bacteriocin produced by *Bacillus subtilis* SN7. LWT-Food Science and Technology. 2018;**87**:8-15

[14] Eom JS, Lee SY, Choi HS. Bacillus subtilis HJ18-4 from traditional fermented soybean food inhibits *Bacillus cereus* growth and toxin-related genes. Journal of Food Science. 2014;**79**(11): M2279-M2287

[15] Park NY, Lee SY, Kim JY, Choi HS. Quality characteristics of buckwheat Soksungjang manufactured by *Bacillus subtilis* HJ18-4. Korean Journal of Food Preservation. 2013;**20**(5):699-704

[16] Eom JS, Song J, Choi HS. Protective effects of a novel probiotic strain

of lactobacillus plantarum JSA22 from traditional fermented soybean food against infection by *Salmonella enterica* serovar Typhimurium. Journal Microbioly Biotechnoly. 2015;**25**(4):479-491

[17] Choi JH, Kim MJ, Cha YS. Cheonggukjang, a soybean paste fermented with B. licheniformis-67 prevents weight gain and improves glycemic control in high fat diet induced obese mice. Journal of Clinical Biochemistry and Nutrition. 2016;**59**(1):31-38

[18] Shin D, Jeong D. Korean traditional fermented soybean products: Jang. Journal of Ethnic Foods. 2015;**2**(1):2-7

[19] Lee JJ, Lee DS, Kim HB. Fermentation patterns of Chungkookjang and Kanjang by *Bacillus licheniformis* B1. The Korean Journal of Microbiology. 1999;**35**(4):296-301

[20] Su CL, Wu CJ, Chen FN, Wang BJ, Sheu SR, Won SJ. Supernatant of bacterial fermented soybean induces apoptosis of human hepatocellular carcinoma Hep 3B cells via activation of caspase 8 and mitochondria. Food and Chemical Toxicology. 2007;**45**(2):303-314

[21] Nakajima N, Nozaki N, Ishihara K, Ishikawa A, Tsuji H. Analysis of isoflavone content in tempeh, a fermented soybean, and preparation of a new isoflavone-enriched tempeh. Journal of Bioscience and Bioengineering. 2005;**100**(6):685-687

[22] Kwon DY, Jang JS, Lee JE, Kim YS, Shin DH, Park S. The isoflavonoid aglycone-rich fractions of Chungkookjang, fermented unsalted soybeans, enhance insulin signaling and peroxisome proliferator-activated receptor-γ activity in vitro. BioFactors. 2006;**26**(4):245-258

[23] Lee SJ, Rim HK, Jung JY, An HJ, Shin JS, Cho CW, et al. Immunostimulatory activity of polysaccharides from Cheonggukjang. Food and Chemical Toxicology. 2013;**59**:476-484

[24] Nout R. Quality, safety, biofunctionality and fermentation control in soya. In: Advances in Fermented Foods and Beverages. Woodhead Publishing Series in Food Science, Technology and Nutrition, Korea. 2015. pp. 409-434

[25] Onda T, Yanagida F, Tsuji M, Shinohara T, Yokotsuka K. Production and purification of a bacteriocin peptide produced by Lactococcus sp. strain GM005, isolated from miso-paste. International Journal of Food Microbiology. 2003;**87**(1-2):153-159

[26] Zhang JH, Tatsumi E, Fan JF, Li LT. Chemical components of Aspergillus-type Douchi, a Chinese traditional fermented soybean product, change during the fermentation process. International Journal of Food Science and Technology. 2007;**42**(3):263-268

[27] Chen T, Jiang S, Xiong S, Wang M, Zhu D, Wei H. Application of denaturing gradient gel electrophoresis to microbial diversity analysis in Chinese Douchi. Journal of the Science of Food and Agriculture. 2012;**92**(10):2171-2176

[28] Wang D, Wang LJ, Zhu FX, Zhu JY, Chen XD, Zou L, et al. In vitro and in vivo studies on the antioxidant activities of the aqueous extracts of Douchi (a traditional Chinese salt-fermented soybean food). Food Chemistry. 2008;**107**(4):1421-1428

[29] Iwaniak A, Minkiewicz P, Darewicz M. Food-originating ACE inhibitors, including antihypertensive peptides, as preventive food components in blood pressure reduction. Comprehensive Reviews in Food Science and Food Safety. 2014;**13**(2):114-134

[30] Wei Q, Chang SK. Characteristics of fermented natto products as

affected by soybean cultivars. Journal of Food Processing and Preservation. 2004;**28**(4):251-273

[31] Sarkar P, Tamang JP, Cook PE, Owens J. Kinema—A traditional soybean fermented food: Proximate composition and microflora. Food Microbiology. 1994;**11**(1):47-55

[32] Sarkar PK, Hasenack B, Nout MJR. Diversity and functionality of Bacillus and related genera isolated from spontaneously fermented soybeans (Indian Kinema) and locust beans (African Soumbala). International Journal of Food Microbiology. 2002;**77**(3):175-186

[33] Cheng AC, Lin HL, Shiu YL, Tyan YC, Liu CH. Isolation and characterization of antimicrobial peptides derived from *Bacillus subtilis* E20-fermented soybean meal and its use for preventing Vibrio infection in shrimp aquaculture. Fish & Shellfish Immunology. 2017;**67**:270-279

[34] Wei LS. Domestic soybean consumption in Asia. In: Soybean Processing for Food Uses. University of Illinois: Urbana-Champaign, INTSOY; 1991. pp. 162-183

[35] De Reu JC, Linssen VA, Rombouts FM, Nout MR. Consistency, polysaccharidase activities and non-starch polysaccharides content of soya beans during tempe fermentation. Journal of the Science of Food and Agriculture. 1997;**73**(3):357-363

[36] Moreno MRF, Leisner JJ, Tee LK, Ley C, Radu S, Rusul G, et al. Microbial analysis of Malaysian tempeh, and characterization of two bacteriocins produced by isolates of *Enterococcus faecium*. Journal of Applied Microbiology. 2002;**92**(1):147-157

[37] Kim HW, Kim KM, Ko EJ, Lee SK, He S, Song KB, et al. Development of antimicrobial edible film from defatted soybean meal fermented by *Bacillus subtilis*. Journal of Microbiology and Biotechnology. 2004;**14**(6):1303-1309

[38] Kim HW, Ko EJ, Ha SD, Song KB, Park SK, Chung DH, et al. Physical, mechanical, and antimicrobial properties of edible film produced from defatted soybean meal fermented by *Bacillus subtilis*. Journal of Microbiology and Biotechnology. 2005;**15**(4):815-822

[39] Baruzzi F, Quintieri L, Morea M, Caputo L. Antimicrobial compounds produced by *Bacillus* spp. and applications in food. Science against Microbial Pathogens: Communicating Current Research and Technological Advances. 2011;**2**:1102-1111

[40] Téllez GA, Castaño JC. Péptidos antimicrobianos. Infection. 2010;**14**(1):55-67

[41] Wimley WC. Describing the mechanism of antimicrobial peptide action with the interfacial activity model. ACS Chemical Biology. 2010;**5**(10):905-917

[42] Li Y, Xiang Q, Zhang Q, Huang Y, Su Z. Overview on the recent study of antimicrobial peptides: Origins, functions, relative mechanisms and application. Peptides. 2012;**37**(2):207-215

[43] Ruiz JC, Segura-Campos MR, Betancur-Ancona D, Chel-Guerrero L. Proteínas y péptidos biológicamente activos con potencial nutracéutico. Omnia Science Monographs. 2013; 11-27

[44] Torrent M, Sánchez-Chardi A, Nogués MV, Boix E. Assessment of antimicrobial compounds by microscopy techniques. In: Microscopy: Science, Technology, Applications and Education. Formatex, Spain. 2010. pp. 1115-1126

[45] Tomasinsig L, Skerlavaj B, Scarsini M, Guida F, Piccinini R, Tossi A,

et al. Comparative activity and mechanism of action of three types of bovine antimicrobial peptides against pathogenic Prototheca spp. Journal of Peptide Science. 2012;**18**(2):105-113

[46] Wang CT, Ji BP, Li B, Nout MJR, Li PL, Ji H, et al. Purification and characterization of a fibrinolytic enzyme of *Bacillus subtilis* DC33, isolated from Chinese traditional Douchi. Journal of Industrial Microbiology and Biotechnology. 2006;**33**:750-758

[47] Nout MJR, Bakshi D, Sarkar PK. Microbiological safety of kinema, a fermented soya bean food. Food Control. 1998;**9**(6):357-362

[48] Donkor ON, Shah NP. Production of β-glucosidase and hydrolysis of isoflavone phytoestrogens by *Lactobacillus acidophilus*, *Bifidobacterium lactis*, and *Lactobacillus casei* in soymilk. Journal of Food Science. 2008;**73**(1):M15-M20

[49] Yesillik S, Yildirim N, Dikici A, Yildiz A. Antibacterial effects of some fermented commercial and homemade dairy products and 0.9% lactic acid against selected foodborne pathogens. Asian Journal of Animal Veterinary Advances. 2011;**6**(2):189-195

[50] Chou CC, Hou JW. Growth of bifidobacteria in soymilk and their survival in the fermented soymilk drink during storage. International Journal of Food Microbiology. 2000;**56**(2-3):113-121

[51] El-Sayed EM, El-Zeini HM, Hafez SA, Saleh FA. Antibacterial activity of probiotic yoghurt and soy-yoghurt against *Escherichia coli* and *Staphylococcus aureus*. Journal of Nutrition and Food Sciences. 2014;**4**(303):2

[52] De Vuyst L, Leroy F. Bacteriocins from lactic acid bacteria: Production, purification, and food applications. Journal of Molecular Microbiology and Biotechnology. 2007;**13**(4):194-199

[53] Biswas K, Upadhayay S, Rapsang GF, Joshi SR. Antibacterial and synergistic activity against β-lactamase-producing nosocomial Bacteria by Bacteriocin of LAB isolated from lesser known traditionally fermented products of India. HAYATI Journal of Biosciences. 2017;**24**(2):87-95

[54] Lee SG, Chang HC. Assessment of *Bacillus subtilis* SN7 as a starter culture for Cheonggukjang, a Korean traditional fermented soybean food, and its capability to control *Bacillus cereus* in Cheonggukjang. Food Control. 2017;**73**:946-953

[55] Cotter PD, Hill C, Ross RP. Bacteriocins: Developing innate immunity for food. Nature Reviews Microbiology. 2005;**3**(10):777-788

[56] Gálvez A, Abriouel H, López RL, Omar NB. Bacteriocin-based strategies for food biopreservation. International Journal of Food Microbiology. 2007;**120**(1-2):51-70

[57] Lee MH, Lee J, Nam YD, Lee JS, Seo MJ, Yi SH. Characterization of antimicrobial lipopeptides produced by *Bacillus* sp. LM7 isolated from chungkookjang, a Korean traditional fermented soybean food. International Journal of Food Microbiology. 2016;**221**:12-18

[58] Ongena M, Jacques P. Bacillus lipopeptides: Versatile weapons for plant disease biocontrol. Trends in Microbiology. 2008;**16**(3):115-125

[59] Shaligram NS, Singhal RS. Surfactin—A review on biosynthesis, fermentation, purification and applications. Food Technology and Biotechnology. 2010;**48**(2):119-134

[60] Lee NK, Yeo IC, Park JW, Kang BS, Hahm YT. Isolation and characterization

of a novel analyte from *Bacillus subtilis* SC-8 antagonistic to *Bacillus cereus*. Journal of Bioscience and Bioengineering. 2010;**110**(3):298-303

[61] Lee NK, Yeo IC, Park JW, Hahm YT. Growth inhibition and induction of stress protein, Gro EL, of *Bacillus cereus* exposed to antibacterial peptide isolated from *Bacillus subtilis* SC-8. Applied Biochemistry and Biotechnology. 2011;**165**(1):235-242

[62] Yeo IC, Lee NK, Hahm YT. Genome sequencing of *Bacillus subtilis* SC-8, antagonistic to the *Bacillus cereus* group, isolated from traditional Korean fermented-soybean food. Journal of Bacteriology. 2012;**194**(2):536-537

[63] Jasim B, Sreelakshmi KS, Mathew J, Radhakrishnan EK. Surfactin, iturin, and fengycin biosynthesis by *Endophytic bacillus* sp. from *Bacopa monnieri*. Microbial Ecology. 2016;**72**(1):106-119

[64] Banat IM, Franzetti A, Gandolfi I, Bestetti G, Martinotti MG, Fracchia L, et al. Microbial biosurfactants production, applications and future potential. Applied Microbiology and Biotechnology. 2010;**87**(2):427-444

[65] Sun L, Lu Z, Bie X, Lu F, Yang S. Isolation and characterization of a co-producer of fengycins and surfactins, endophytic *Bacillus amyloliquefaciens* ES-2, from *Scutellaria baicalensis* Georgi. World Journal of Microbiology and Biotechnology. 2006;**22**(12):1259-1266

[66] Zhang N, Pu Y, Sun L, Wang Y, Deng Q, Xu D, et al. Modeling the effects of different conditions on the inhibitory activity of antimicrobial lipopeptide (AMPNT-6) against *Staphylococcus aureus* growth and enterotoxin production in shrimp meat. Aquaculture International. 2017;**25**(1):57-70

[67] Xu D, Wang Y, Sun L, Liu H, Li J. Inhibitory activity of a novel antibacterial peptide AMPNT-6 from *Bacillus subtilis* against *Vibrio parahaemolyticus* in shrimp. Food Control. 2013;**30**(1):58-61

[68] Chen X, Stewart PS. Biofilm removal caused by chemical treatments. Water Research. 2000;**34**(17):4229-4233

[69] Croxatto A, Lauritz J, Chen C, Milton DL. Vibrio anguillarum colonization of rainbow trout integument requires a DNA locus involved in exopolysaccharide transport and biosynthesis. Environmental Microbiology. 2007;**9**(2):370-382

[70] Branda SS, Vik Å, Friedman L, Kolter R. Biofilms: The matrix revisited. Trends in Microbiology. 2005;**13**(1):20-26

[71] Deng Q, Pu Y, Sun L, Wang Y, Liu Y, Wang R, et al. Antimicrobial peptide AMPNT-6 from Bacillus subtilis inhibits biofilm formation by *Shewanella putrefaciens* and disrupts its preformed biofilms on both abiotic and shrimp shell surfaces. Food Research International. 2017;**102**:8-13

[72] Gabriel AA, Nakano H. Inactivation of *Salmonella*, *E. coli* and *Listeria monocytogenes* in phosphate-buffered saline and apple juice by ultraviolet and heat treatments. Food Control. 2009;**20**(4):443-446

[73] O'Connor NK, Hudson AS, Cobb SL, O'Neil D, Robertson J, Duncan V, et al. Novel fluorinated lipopeptides from Bacillus sp. CS93 via precursor-directed biosynthesis. Amino Acids. 2014;**46**(12):2745-2752

[74] Mandal SM, Sharma S, Pinnaka AK, Kumari A, Korpole S. Isolation and characterization of diverse antimicrobial lipopeptides produced by Citrobacter and Enterobacter. BMC Microbiology. 2013;**13**(1):152

[75] Lee SY, Kim JY, Baek SY, Yeo SH, Koo BS, Park HY, et al. Isolation

and characterization of oligotrophic strains with high enzyme activity from buckwheat sokseongjang. Korean Journal of Food Science and Technology. 2011;**43**(6):735-741

[76] Sumi CD, Yang BW, Yeo IC, Hahm YT. Antimicrobial peptides of the genus Bacillus: A new era for antibiotics. Canadian Journal of Microbiology. 2014;**61**(2):93-103

[77] Yeo IC, Lee NK, Cha CJ, Hahm YT. Narrow antagonistic activity of antimicrobial peptide from Bacillus subtilis SCK-2 against *Bacillus cereus*. Journal of Bioscience and Bioengineering. 2011;**112**(4):338-344

[78] Shafi J, Tian H, Ji M. Bacillus species as versatile weapons for plant pathogens: A review. Biotechnology & Biotechnological Equipment. 2017;**31**(3):446-459

[79] Cleveland J, Montville TJ, Nes IF, Chikindas ML. Bacteriocins: Safe, natural antimicrobials for food preservation. International Journal of Food Microbiology. 2001;**71**(1):1-20

[80] De Vuyst L, Moreno MF, Revets H. Screening for enterocins and detection of hemolysin and vancomycin resistance in enterococci of different origins. International Journal of Food Microbiology. 2003;**84**(3):299-318

[81] Gálvez A, Valdivia E, Abriouel H, Camafeita E, Mendez E, Martínez-Bueno M, et al. Isolation and characterization of enterocin EJ97, a bacteriocin produced by *Enterococcus faecalis* EJ97. Archives of Microbiology. 1998;**171**(1):59-65

[82] Herranz C, Casaus P, Mukhopadhyay S, Martınez JM, Rodrıguez JM, Nes IF, et al. *Enterococcus faecium* P21: A strain occurring naturally in dry-fermented sausages producing the class II bacteriocins enterocin A and enterocin B. Food Microbiology. 2001;**18**(2):115-131

[83] Hugas M, Garriga M, Aymerich MT. Functionalty of enterococci in meat products. International Journal of Food Microbiology. 2003;**88**(2-3):223-233

[84] Leroy F, Moreno MF, De Vuyst L. Enterococcus faecium RZS C5, an interesting bacteriocin producer to be used as a co-culture in food fermentation. International Journal of Food Microbiology. 2003;**88**(2-3):235-240

[85] Nam HR, Ha MS, Lee EJ, Lee YH. Effect of *Enterococcus faecalis* strain PL9003 on adherence and growth of *Helicobacter pylori*. Journal of Microbiology and Biotechnology. 2002;**12**(5):746-752

[86] Tsai CC, Huang LF, Lin CC, Tsen HY. Antagonistic activity against *Helicobacter pylori* infection in vitro by a strain of *Enterococcus faecium* TM39. International Journal of Food Microbiology. 2004;**96**(1):1-12

[87] Yoon MY, Kim YJ, Hwang HJ. Properties and safety aspects of *Enterococcus faecium* strains isolated from Chungkukjang, a fermented soy product. LWT-Food Science and Technology. 2008;**41**(5):925-933

[88] Onda T, Yanagida F, Tsuji M, Ogino S, Shinohara T. Isolation and characterization of the lactic acid bacterial strain GM005 producing a antibacterial substance from miso-paste product. Food Science and Technology Research. 1999;**5**(3):247-250

[89] Gross E, Morell JL. Structure of nisin. Journal of the American Chemical Society. 1971;**93**(18):4634-4635

[90] Shiu YL, Wong SL, Guei WC, Shin YC, Liu CH. Increase in the plant protein ratio in the diet of white shrimp,

Litopenaeus vannamei (Boone), using *Bacillus subtilis* E20-fermented soybean meal as a replacement. Aquaculture Research. 2015;**46**(2):382-394

[91] Cheng W, Chiu CS, Guu YK, Tsai ST, Liu CH. Expression of recombinant phytase of *Bacillus subtilis* E 20 in *Escherichia coli* HMS 174 and improving the growth performance of white shrimp, *Litopenaeus vannamei*, juveniles by using phytase-pretreated soybean meal-containing diet. Aquaculture Nutrition. 2013;**19**(2):117-127

[92] Cao XH, Liao ZY, Wang CL, Yang WY, Lu MF. Evaluation of a lipopeptide biosurfactant from *Bacillus natto* TK-1 as a potential source of anti-adhesive, antimicrobial and antitumor activities. Brazilian Journal of Microbiology. 2009;**40**(2):373-379

[93] Sheppard JD, Jumarie C, Cooper DG, Laprade R. Ionic channels induced by surfactin in planar lipid bilayer membranes. Biochimica et Biophysica Acta (BBA)—Biomembranes. 1991;**1064**(1):13-23

[94] Kim PI, Bai H, Bai D, Chae H, Chung S, Kim Y, et al. Purification and characterization of a lipopeptide produced by *Bacillus thuringiensis* CMB26. Journal of Applied Microbiology. 2004;**97**(5):942-949

[95] Dhayakaran RPA, Neethirajan S, Xue J, Shi J. Characterization of antimicrobial efficacy of soy isoflavones against pathogenic biofilms. LWT-Food Science and Technology. 2015;**63**(2):859-865

[96] Dhayakaran R, Neethirajan S, Weng X. Investigation of the antimicrobial activity of soy peptides by developing a high throughput drug screening assay. Biochemistry and Biophysics Reports. 2016;**6**:149-157

[97] McClean S, Beggs LB, Welch RW. Antimicrobial activity of antihypertensive food-derived peptides and selected alanine analogues. Food Chemistry. 2014;**146**:443-447

[98] Capriotti AL, Caruso G, Cavaliere C, Samperi R, Ventura S, Chiozzi RZ, et al. Identification of potential bioactive peptides generated by simulated gastrointestinal digestion of soybean seeds and soy milk proteins. Journal of Food Composition and Analysis. 2015;**44**:205-213

[99] Rashed NA, Mac Donald MH, Matthews BF. Protease inhibitor expression in soybean roots exhibiting susceptible and resistant interactions with soybean cyst nematode. Journal of Nematology. 2008;**40**(2):138